SUMMABILITY THROUGH FUNCTIONAL ANALYSIS

NORTH-HOLLAND
MATHEMATICS STUDIES **85**

Notas de Matemática (91)
Editor: Leopoldo Nachbin

Centro Brasileiro de Pesquisas Físicas
and University of Rochester

Summability through Functional Analysis

ALBERT WILANSKY
Lehigh University

1984

NORTH-HOLLAND – AMSTERDAM • NEW YORK • OXFORD

ISBN: 0 444 86840 2

Publishers:

ELSEVIER SCIENCE PUBLISHERS B.V.
P.O. BOX 1991
1000 BZ AMSTERDAM
THE NETHERLANDS

Sole distributors for the U.S.A. and Canada:

ELSEVIER SCIENCE PUBLISHING COMPANY, INC.
52 VANDERBILT AVENUE
NEW YORK, N.Y. 10017

Library of Congress Cataloging in Publication Data

Wilansky, Albert.
Summability through functional analysis.

(North-Holland mathematics studies ; 85) (Notas de matemática ; 91)
Bibliography: p.
Includes index.
1. Functional analysis. 2. Summability theory.
I. Title. II. Series. III. Series: Notas de matemática (Amsterdam, Netherlands) ; 91.
QA320.W56 1984 515.7 83-25398
ISBN 0-444-86840-2

PRINTED IN THE NETHERLANDS

To Carole, Eleanor, Johnny,
Kathy, Laura, Leslie, Michael

Unique members of a unique family

Herrn Karl Zeller zum 60. Geburtstag
am 28. Dezember 1984 gewidmet.

PREFACE

Summability is an extremely fruitful area for the application of functional analysis; this book could be used as a source for such applications. Those parts of summability which have only "hard" (= classical) proofs are omitted; the theorems given all have "soft" (= functional analytic) proofs. (There are a few exceptions.) Here is an incomplete list of topics *not* covered: normal spaces, perfect spaces, ordered spaces, classical Tauberian theorems, summability of Fourier series, absolute and strong summability, summability factors, discussion of dozens of special methods. These are all interesting and valuable topics but do not fall within our scope. The monographs [55], [88] are recommended for material not in this book.

I have found summability through functional analysis to be a most entertaining subject — full of interesting things to work on with amusing and challenging problems at every stage. Further, it rewards the expert in functional analysis with delightful short proofs. When I was in graduate school it was customary to assign the necessity of the Silverman-Toeplitz conditions (1.3.6) as a homework problem. The best and bravest of us found the right construction — those who did can really appreciate the soft proof given here; some of us can even remember the sensation when it first appeared in Banach's book [2]. In connection with this I must take exception to a remark of Ivor Maddox in [44A], p. 162. He indicates that functional analysis smooths the path of proof but does not obtain the actual results to be proved — the methods giving existence but not construction. On the contrary, as in Chapter 8, the results themselves appear while the classical writers had to guess what they were and then prove them.

SOURCES.

I have given references and attributions for some of the results. The reader will not go far wrong to assume that every unattributed theorem outside of Chapter 2 is due to Karl Zeller.

NOTICE

Coregular and conull have special meanings in Chapters 1-8 and 17-19. In those chapters these spaces are conservative. *In Chapters 9-16* they are assumed to have the weaker property variational semiconservative, written vsc throughout.

ACKNOWLEDGMENTS

A. K. Snyder and W. H. Ruckle helped me over dozens of sticky points. A group of Lehigh University graduate students read over the manuscript and made many suggestions and corrections. They also attended a seminar on this material led by Professor R. M. DeVos of Villanova University. The group consisted of Deb Frantz, Abdullah Hakawati and Matt Schaffer. They were joined for a time by Jeff Connor of Kent State University who also proof-read part of the typescript. I thank all these friends for their help.

Judy Arroyo undertook the difficult task of typing camera-ready copy. Her enthusiasm and proprietary attitude were indispensable. As to the quality of her work: "Si monumentum requiris, circumspice."

CONTENTS

CHAPTER 1
MATRICES

1.0. FUNCTIONAL ANALYSIS

Reference: [80], Chapter 3.

1. A *Banach space* is a complete normed vector space X; X' is the dual Banach space of continuous linear functionals (i.e. scalar valued functions) on X with

$$\|f\| = \sup\{|f(x)| : \|x\| \leq 1\}; \quad f \in X' \quad \text{iff} \quad \|f\| < \infty;$$
$$|f(x)| \leq \|f\| \cdot \|x\| .$$

2. c_o (the null sequences), c (the convergent sequences) and ℓ^∞ (the bounded sequences) are Banach spaces with $\|x\|_\infty = \sup|x_n|$. Each is a closed subspace of the next. ℓ (the absolutely convergent series) is a Banach space with $\|x\|_1 = \Sigma\, |x_k|$. $f \in c_o'$ iff $f(x) = ax$ with $a \in \ell$, (see 1.2.1 for ax), $\|f\| = \|a\|_1$; $f \in c'$ iff $f(x) = \chi \lim x + ax$ with $a \in \ell$, $\chi(f) = f(1) - \Sigma\, f(\delta^k)$ (1.3.5), $f \in \ell'$ iff $f(x) = ax$ with $a \in \ell^\infty$, $\|f\| = \|a\|_\infty$. [80], Examples 2.3.5, 2.3.6, #2-3-3.

3. *UNIFORM BOUNDEDNESS*. Let $\{f_n\} \subset X'$, X a Banach space, and suppose that $\{f_n\}$ is pointwise bounded, i.e. $|f_n(x)| < M_x$ for all n,x. Then $\{f_n\}$ is uniformly bounded i.e. $\|f_n\| < M$ for all n. See [80], Theorem 3-3-6.

4. *BANACH-STEINHAUS CLOSURE THEOREM*. With $\{f_n\}$ as in 3 suppose that $f(x) = \lim f_n(x)$ for all $x \in X$. Then $f \in X'$. [80], Theorem 3-3-13. This holds also if X is a locally convex Fréchet space [80], Theorem 9-3-7, Example 9-3-2.

5. *CONVERGENCE LEMMA*. Let $\{f_n\}$ be uniformly bounded (as in 3). Then $\{x:\lim f_n(x) \text{ exists}\}$ and $\{x:f_n(x) \to 0\}$ are closed vector subspaces of X. [80], #3-3-3.

6. A *Banach algebra* is a Banach space which is also an algebra and satisfies $\|xy\| \leq \|x\|\cdot\|y\|$. If X is a Banach algebra with identity and $\|x-1\| < 1$ then x is invertible. [79], 14.2 Fact ii.

7. A topological vector space is a vector space and a topological space such that the vector operations are continuous. Each such space X has a weak topology w which is the smallest vector topology with the same dual. A sequence $\{a^n\} \to 0$ in w iff $f(a^n) \to 0$ for all $f \in X'$. [80], Example 4-1-9.

1.1. INTRODUCTION

In the beginning, the idea was conceived that there should be a way to find "sums" for divergent series. One popular procedure was to set $x = 1$ in the identity $(1+x)^{-1} = \Sigma\,(-x)^n$ leading to the mystically satisfying "result" that $1-1+1-1+\ldots = \frac{1}{2}$.

Since it is obviously possible to assign a sum, for example 0, to any divergent series, we abandon this quest and simply look for some type of function $L:S \to K$ where S is some set of sequences and K is the set of scalars - always R (the real numbers) or C (the complex numbers) in this book. Note that "sequences" has been substituted for "series"; this is a mild and unnecessary change which is adopted for convenience. The function L will be required to have certain explicitly stated properties; for example, we usually require S to be a vector space which includes all the convergent sequences and L to be linear and such that $L(x) = \lim x_n$ whenever $x = \{x_n\}$ is convergent. Then if S contains a divergent sequence x, the number $L(x)$ will be a "limit of a divergent sequence" in the eldritch sense described above.

It turns out that such functions may be very practical. See §19.4.

1.2. NOTATION.

There is an Index of Symbols in the book which refers to the place where each symbol is defined.

1. A sequence $\{x_n\}$, $n = 1,2,\ldots,$ of scalars will be written x. *The number* $\Sigma\, x_k y_k$ *will be denoted by* xy. Given a matrix $A = (a_{nk})$, $n,k = 1,2,\ldots$ of scalars and a sequence x, we write $y = Ax$ to mean that for each $n, y_n = (Ax)_n = \Sigma_{k=1}^{\infty} a_{nk}x_k$; each of these series being assumed convergent.

Let $\omega_A = \{x{:}Ax$ is defined$\}$; this is called the *domain* of A. If A is *row-finite* (i.e. each row of A lies in ϕ, the set of all finitely non-zero sequences) then $\omega_A = \omega$, the set of all sequences; it is a simple exercise to check that the converse is also true.

2. Let $c_A = \{x{:}Ax \in c\}$ where c is the set of convergent sequences. This is called the *convergence domain* of A. More generally we define $S_A = \{x{:}Ax \in S\}$ where S is any subset of ω; this is consistent with the meaning of ω_A as just given. We emphasize that $Ax \in S$ always implies that Ax exists, i.e. $S_A \subset \omega_A$ for every S.

By a historical accident, sequences in c_A are called *A-summable* instead of the more reasonable A-limitable.

3. For $x \in c_A$ let $\lim_A x = \lim(Ax)_n$, thus defining $\lim_A{:}c_A \to K$. Finally, a matrix A is called *conservative* if $c_A \supset c$ (i.e. $Ax \in c$ whenever $x \in c$), *multiplicative* m if

$\lim_A x = m\cdot\lim x$ for $x \in c$, and *regular* if it is multiplicative 1.

4. EXAMPLE. The identity matrix I is regular since $Ix = x$; also $\omega_I = \omega$, $c_I = c$.

5. EXAMPLE. Let Q be the matrix given by $(Qx)_1 = x_1$, $(Qx)_n = (x_{n-1}+x_n)/2$ for $n > 1$. Then Q is regular and sums the divergent sequence $\{(-1)^n\}$.

6. EXAMPLE. For $A = 0$, $c_A = \omega$ while at the opposite extreme one can construct A with $c_A = \{0\}$ by taking $Ax = (x_1, 0, x_1, x_2, 0, x_1, x_2, x_3, 0, \ldots)$.

7. DEFINITION. *Let X, Y be sets of sequences. Then $(X:Y)$ is the set of matrices A such that $Ax \in Y$ for all $x \in X$.*

Thus $A \in (\omega_A:\omega)$ always; $A \in (c:c)$ iff A is conservative.

8. DEFINITION. *For a complex number z, $sgn\ z = |z|/z$ if $z \neq 0$, $sgn\ 0 = 1$. Thus $|sgn\ z| = 1$ and $z\ sgn\ z = |z|$ for all z.*

9. Here is a version of *Abel's identity* that I find very easy to remember. Suppose $v_0 = u_{n+1} = 0$, then

$$\Sigma_{k=1}^n u_k(v_k \pm v_{k-1}) = \Sigma_{k=1}^n (u_k \pm u_{k+1})v_k.$$

(Take $+$ in both or $-$ in both). Then, for example, if $u_{n+1} \neq 0$, one merely adds $\pm u_{n+1}v_n$ to the left side.

10. *Abel's Inequality. Suppose $1 \geq u_1 \geq u_2 \ldots \geq u_n \geq 0$. Then $|\Sigma_{k=1}^n u_k e_k| \leq \max\{|\Sigma_{k=1}^r e_k| : 1 \leq r \leq n\}$.*

Let $v_k = \Sigma_{i=1}^k e_i$ and set $u_{n+1} = 0$. Then by Abel's identity, $|\Sigma u_k e_k| = |\Sigma u_k(v_k - v_{k-1})| = |\Sigma (u_k - u_{k+1})v_k| \leq \max|v_k|\cdot\Sigma (u_k - u_{k+1}) \leq \max |v_k|$.

1.3. CONSERVATIVE AND REGULAR

The main purpose of this section is to characterize members of (c:c) (1.2.7). A unified approach to this problem for many pairs of spaces is given in Chapter 8.

1. We now refer to the spaces c_o, ℓ, ℓ^∞ and their norms as given in 1.0.2.

2. THEOREM. *For a matrix A, these are equivalent:* *(i)* $\omega_A \supset \ell^\infty$, *(ii)* $\omega_A \supset c_o$, *(iii)* $\Sigma_k|a_{nk}| < \infty$ *for each n.*

That (iii) implies (i) and (i) implies (ii) are trivial. Assuming (ii) let a be any row of A and $u_m(x) = \Sigma_{k=1}^m a_k x_k$, defining $u_m \in c_o'$ with $\|u_m\| = \Sigma_{k=1}^m |a_k|$ (1.0.2). The sequence u is pointwise convergent, hence pointwise bounded so norm bounded (1.0.3). This yields (iii).

3. THEOREM. *For a matrix A, these are equivalent:* *(i)* $A \in (\ell^\infty:\ell^\infty)$, *(ii)* $A \in (c_o:\ell^\infty)$, *(iii)* $\|A\| < \infty$ *where* $\|A\| = sup_n \Sigma_k|a_{nk}|$.

(iii) implies (i): If x is bounded, $|(Ax)_n| \leq \|A\|\cdot\|x\|_\infty$.

(ii) implies (iii): Define $u_n \in c_o'$ by $u_n(x) = (Ax)_n$ so that $\|u_n\| = \Sigma|a_{nk}|$ (1.0.2). The sequence u is pointwise bounded, hence norm bounded (1.0.3) and the result follows.

4. The set of matrices of finite norm (see Theorem 3(iii)) is denoted by Φ. Thus $\Phi = (\ell^\infty:\ell^\infty) = (c_o:\ell^\infty)$ by Theorem 3. It is easy to check that $\|A\|$ is the usual norm of the map $x \to Ax$ from c_o or ℓ^∞ to ℓ^∞.

5. NOTATION (very Important!). For a matrix A, $a_k = lim_n a_{nk}$ for each k, provided these (column) limits exist. Similarly b_k is defined for a matrix B. For each $k = 1,2,\ldots;\delta^k$ is the

sequence whose only non-zero term is a 1 in the kth place; 1 is either the integer or the sequence all of whose terms are 1 (according to context). It is crucial that $a_k = \lim_A \delta^k$.

6. THEOREM. *A matrix* A *is conservative iff* (i) $A \in \Phi$, (ii) a_k *exists for each* k, (iii) $1 \in c_A$. *The conditions* (i), (ii) *are necessary and sufficient that* $c_A \supset c_o$.

Let A be conservative; (i) is a special case of Theorem 3; (ii) and (iii) hold because $c_A \supset c$ and, for example, a_k exists iff $\delta^k \in c_A$. If, conversely, the three conditions hold, let $F(x) = Ax$ define $F:\ell^\infty \to \ell^\infty$, a continuous linear map, by (i). Then $U = F^{-1}[c]$ is a closed vector subspace of ℓ^∞ since c is. Now (ii) and (iii) imply that each $\delta^k \in U$ and $1 \in U$. Since c is the smallest closed vector subspace of ℓ^∞ which contains all these sequences, it follows that $U \supset c$ and so A is conservative. The second part of the theorem has the same proof with all considerations involving 1 omitted, and c replaced by c_o.

7. THEOREM. *If* $A \in \Phi$ *and each* a_k *exists* (*i.e.* $\omega_A \supset \phi$) *then* $a \in \ell$.

For each m, $\Sigma_{k=1}^m |a_k| = \lim_n \Sigma_{k=1}^m |a_{nk}| \leq \|A\|$.

It is not true that $\Sigma\, a_{nk}$ must be uniformly convergent, as the identity matrix shows.

8. THEOREM. *Let* A *be conservative and* $x \in c$. *Then* $\lim_A x = \chi \cdot \lim x + ax$ *where* $\chi = \chi(A) = \lim_A 1 - \Sigma\, a_k$, ax *is as in 1.2.1.*

The Banach-Steinhaus Theorem (1.0.4) shows that $\lim_A \in c'$ and the formula follows from 1.0.2.

9. THEOREM. *A matrix is regular iff* $\|A\| < \infty$, $a_k = 0$ *for each* k, *and* $\lim_A 1 = 1$.

Necessity is by Theorem 6 and the fact that $a_k = \lim_A \delta^k$. Sufficiency is by Theorems 6 and 8.

The conditions of Theorem 9 are called the (L.L.) Silverman-(O.) Toeplitz conditions. Their discovery in 1911 marked the beginning of the general theory — as opposed to the study of specific matrices. The original proof by elementary analysis is an excellent — and difficult — exercise for graduate students. The "soft" proof given here is due to S. Banach in 1932. It marks the beginning of the functional analysis approach.

10. EXAMPLE. The Cesàro matrix denoted by $(C,1)$ is the matrix A such that $(Ax)_n$ is the average of $x_1, x_2, \ldots, x_n$. In symbols, $a_{nk} = 1/n$ for $k = 1,2,\ldots,n$, and for $k > n$, $a_{nk} = 0$. Every baseball player knows that $(C,1)$ is regular; for example, averaging 2 hits in 5 at bats over a long season will result in a batting average of .400 no matter how badly or how well the season started. Mere mathematicians observe that $(C,1)$ obeys the conditions of Theorem 9.

11. THEOREM. *Let* $A \in \Phi$. *Then* $c_A \cap \ell^\infty$ *and* $c_A^o \cap \ell^\infty$ *are closed subspaces of* ℓ^∞.

Here $c_A^o = \{x : Ax \in c_o\}$. Let $u_n(x) = (Ax)_n$ so that $u_n \in (\ell^\infty)'$, indeed $\|u_n\| \le \|A\|$. The result follows by 1.0.5.

1.4. ASSOCIATIVITY

1. DEFINITION. *A matrix* A *is called triangular if* $a_{nk} = 0$ *for* $k > n$, *it is called a triangle if it is triangular and* $a_{nn} \neq 0$ *for all* n.

2. *The expressions* $t(Ax)$ *and* $(tA)x$ *arise often in summability.* Here t, x are sequences and A is a matrix;

$t(Ax) = \Sigma\, t_n(Ax)_n = \Sigma_n\Sigma_k\, t_n a_{nk} x_k$, $(tA)x = \Sigma\,(tA)_k x_k = \Sigma_k\Sigma_n\, t_n a_{nk} x_k$. *They may be different even if* $t \in \ell$, A *is a regular triangle,* $x \in c_A$ *and both numbers exist.* For example let $t_n = 2^{-n}$, $(Ax)_n = 2x_{n-1} - x_n$, so that $tA = 0$. It is very easy to find x such that $t(Ax) \neq 0$; indeed one can solve $y = Ax$ for x by an easy induction for arbitrary y, e.g. $y = \delta^1$.

3. Valuable results ensue whenever it is known that equality holds for these two expressions. A key step in several arguments is just the appeal to this. The entry "associativity" in the index may be consulted.

4. THEOREM. *The equality* $t(Ax) = (tA)x$ *holds if (i)* $t \in \phi$ *and* $x \in \omega_A$ *or (ii)* $t \in \ell$, $A \in \Phi$, $x \in \ell^\infty$. The first part is obvious, involving only the adding together of finitely many convergent series. If (ii) holds, $\Sigma\Sigma|t_n a_{nk} x_k| \leq \|t\|_1 \|A\| \cdot \|x\|_\infty$ and interchange of summation for an absolutely convergent series is allowed.

5. COROLLARY. *The set* Φ *and the set of row-finite matrices have associative multiplication.*

For $A \in \Phi$, the rows of A are in ℓ, the columns in ℓ^∞ so Theorem 4 (ii) applies. The other part follows from the observation that (i) of Theorem 4 holds for all x if A is row-finite.

6. EXAMPLE. $B \circ A \neq BA$, *indeed* $B \circ A$ *is not a matrix map.* The example is nothing but the ordinary telescopic series of our childhood. Let $(Ax)_n = x_n - x_{n-1}$ (Convention: $x_0 = 0$). Then $1(Ax) = \Sigma(Ax)_n = \lim x$ for $x \in c$. Now let B be the matrix whose first row is 1, other rows 0. Then $B(Ax) = (\lim x, 0, 0, ..)$ in particular $B \circ A \neq 0$, while $BA = 0$. To see that $B \circ A : c \to c$ is not given by a matrix it is sufficient to observe that it vanishes

on ϕ; the terms of a matrix are determined by how it maps ϕ since $m_{nk} = (M\delta^k)_n$ for any matrix M. More is true: *if* $B \circ A$ *were a matrix it would have to be the matrix* BA since $[B(A\delta^k)]$ $= (BA)_{nk}$ by an easy calculation i.e. $B \circ A = BA$ on ϕ.

Examples of this type are available whenever $t(Ax) \neq (tA)x$. The example in Remark 2 leads to a regular A and conservative B.

7. In cases like Corollary 5 in which $(BA)x = B(Ax)$ for matrices A, B and x in the space involved, $B \circ A = BA$ by definition.

8. If matrix multiplication is associative, as in Corollary 5, a matrix has an inverse matrix (which is automatically unique) iff it has a unique right (or left) inverse. *A triangle A has a unique right inverse B:* it may be directly computed by induction and is unique since A is one to one. Moreover B is a triangle and $b_{nn} = 1/a_{nn}$. Also $BA = I$ since $A(I-BA) = 0$ by Corollary 5; thus $B = A^{-1}$. *But A may have another left inverse:* just add t to any row of B in the example of Remark 2. See also 1.8.5.

9. Examples 6 and 8 show that it is dangerous to denote the map $x \to Ax$ by A. In 1.3.6 we denoted it by F for this reason. Then F^{-1} could be written without worrying about the ambiguous A^{-1}. In 5.4.8 we shall see a case in which A^{-1} exists but does not give the inverse map.

1.5. THE ALGEBRAS Γ AND Δ

1. DEFINITION. *The set of conservative matrices is denoted by* Γ.

Thus $\Gamma = (c:c)$. By 1.3.6, $\Gamma \subset \Phi$.

2\. LEMMA. *Let* $S_k = \{A \in \Phi: \lim_n a_{nk}$ *exists*$\}$. *Then each* S_k *is a closed vector subspace of* Φ. Let $u_n(A) = a_{nk}$, defining $u_n \in \Phi'$ since $|u_n(A)| \leq \|A\|$. The result follows by 1.0.5.

3\. LEMMA. *Let* $S_0 = \{A \in \Phi: \lim_n \Sigma_k a_{nk}$ *exists*$\}$. *Then* S_0 *is a closed vector subspace of* Φ.

Let $u_n(A) = \Sigma a_{nk}$ so that $|u_n(A)| \leq \|A\|$. The result follows as in Lemma 2.

4\. THEOREM. Γ *is a closed subalgebra of* Φ, *hence it is a Banach algebra.*

By 1.3.6, $\Gamma = \cap\{S_k : k = 0,1,2,\ldots\}$ hence is a closed vector subspace of Φ by Lemmas 2, 3. If $A,B \in \Gamma$, let $x \in c$. Then $(BA)x = B(Ax) \in c$ using 1.4.4 (ii) with t taken to be any row of B; thus $BA \in \Gamma$. It is routine to prove that Φ is a Banach algebra using 1.4.5.

5\. THEOREM. *The set* Δ *of triangular conservative matrices is a closed subalgebra of* Γ. *Hence it is a Banach algebra.*

With $u_{nk}(A) = a_{nk}$, $u_{nk} \in \Gamma'$ and $\Delta = \cap\{u_{nk}^{\perp} : k > n\}$ is closed. It is obviously closed under multiplication.

1.6. COREGULAR AND CONULL MATRICES.

1\. DEFINITION. *A matrix* A *is called conull, respectively, coregular if it is conservative and* $\chi(A) = 0$, *respectively,* $\neq 0$.

2\. The number χ was given in 1.3.8. A regular matrix has $\chi = 1$.

3\. It is possible to define χ for matrices which are not conservative but Definition 1 explicitly excludes such matrices. Beginning in §9.4 we shall relax the assumption that A is conservative, but it will never be entirely dropped i.e. a matrix with $\chi(A) = 0$ will not be called conull unless it has an extra property, namely (until further notice) that it is conservative.

4. The formula $\chi(A) = \lim_m \lim_n \Sigma_{k=m}^{\infty} a_{nk}$ may be proved as follows: the right side is $\lim_m(\lim_n \Sigma_{k=1}^{\infty} a_{nk} - \Sigma_{k=1}^{m-1} a_k)$ $= \lim_n \Sigma_k a_{nk} - \Sigma a_k = \chi(A)$. It follows that $|\chi(A)| \leq \|A\|$ and so χ is a continuous linear function on $\Gamma(1.0.1)$.

1.7. TYPES OF SUMMABILITY THEOREMS.

(a) *GROWTH THEOREMS*

1. THEOREM. *Let A be a triangle, $B = A^{-1}$, $u_n = \Sigma_{k=1}^{n} |b_{nk}|$. Then every $x \in c_A$ satisfies $x_n = O(u_n)$ i.e. $\{x_n/u_n\}$ is bounded.*

From $x = B(Ax)$ (1.4.4 i) follows $|x_n| = |\Sigma\, b_{nk}(Ax)_k| \leq u_n \cdot \|Ax\|_{\infty}$.

2. EXAMPLE. Applying this to $(C,1)$ or Q (1.2.5) yields $x_n = O(n)$. An amusing, perhaps unexpected, result can be deduced, namely, if $\lim(x_n - x_{n+2})$ exists then $x_n - x_{n+1} = O(n)$. The proof is that $\{x_n - x_{n+1}\} \in c_Q$.

(b) *TAUBERIAN THEOREMS*

3. THEOREM. *Let $x \in c_Q$ (1.2.5) and suppose that $\{x_n - x_{n-1}\}$ is convergent. Then x is convergent.*

For $x_n = \frac{1}{2}(x_n + x_{n-1}) + \frac{1}{2}(x_n - x_{n-1})$.

A *Tauberian Theorem* is one in which convergence of a sequence is deduced from convergence of some transform together with a side condition, in this case the assumption that $\{x_n - x_{n-1}\}$ is convergent. The first such theorem was given by A. Tauber in 1897.

(c) *INCLUSION THEOREMS*

4. DEFINITION. *If* $c_B \supset c_A$, *B is said to be stronger than A. If* $c_B = c_A$, *A, B are called equipotent.*

5. THEOREM. *Let A, B be triangles. Then B is stronger than A iff* BA^{-1} *is conservative.*

The proof will make several uses of associativity (1.4.4 i). Necessity: let $x \in c$. Then $A^{-1}x \in c_A \subset c_B$ so $BA^{-1}x \in c$. Sufficiency: let $x \in c_A$. Then $Ax \in c$ so $Bx = (BA^{-1})Ax \in c$ hence $x \in c_B$.

6. EXAMPLE. *(C,1) is stronger than Q* (1.2.5). First note that a typical row of Q^{-1} is $(-1,2,-2,2)$. (This is the 4th row.) The row sums are all 1. This is true because the row sums of Q are all 1. $(Q^{-1}1 = Q^{-1}Q1 = 1)$. It was for the sake of this convenient check on calculation that the first entry of Q was made 1 instead of the more natural 1/2. The row sums of CQ^{-1} are also 1 by a similar argument; all the entries are non-negative and its column limits are 0. Thus CQ^{-1} is regular (1.3.9) and the result follows by Theorem 5.

(d) *BIGNESS THEOREMS.*

7. THEOREM. *Let A be a triangle with* $a_{nn} \to 0$ *(or just* $1/a_{nn}$ *unbounded.) Then A sums an unbounded sequence.*

With $B = A^{-1}$, $\|B\| \geq |b_{nn}| = |1/a_{nn}|$ so $\|B\| = \infty$. Hence there exists $x \in c_o$ with Bx unbounded. (1.3.3). Now $A(Bx) = x \in c_o$ so $Bx \in c_A$.

8. EXAMPLE. *If* A, B *are triangles with* a_{nn}/b_{nn} *unbounded,* A *is not stronger than* B. For $AB^{-1} \notin \Phi$ as in Theorem 7 and the result follows by Theorem 5. In particular Q is not stronger than $(C,1)$ i.e. (Example 6) they are not equipotent.

(e) *CONSISTENCY THEOREMS*

9. DEFINITION. *Matrices* A, B *are called consistent if* $\lim_A x = \lim_B x$ *whenever* $x \in c_A \cap c_B$. *The conditions:* B *is stronger than* A *and consistent with* A *are written* $B \supset A$. *If* $A \supset B \supset A$, A, B *are called equivalent.*

10. THEOREM. *Let* A, B *be triangles. Then* $B \supset A$ *iff* BA^{-1} *is regular.*

Necessity: Let $x \in c$. Using Theorem 5, $\lim BA^{-1}x = \lim_B A^{-1}x = \lim_A A^{-1}x = \lim x$. Sufficiency: Let $x \in c_A \cap c_B = c_A$ by Theorem 5. Then $\lim_B x = \lim_B A^{-1}Ax = \lim Ax$ since BA^{-1} is regular.

For example, $(C,1) \supset Q$ by the argument of Example 6.

11. EXAMPLE. A Tauberian theorem. *If* $x_n - x_{n-1} \to L$ *and* $x_n = o(n)$, *(e.g. if* x *is bounded) then* $L = 0$. Let $(Ax)_n = x_n - x_{n-1}$ $(x_0 = 0)$, $(Bx)_n = x_n/n$. Then $BA^{-1} = (C,1)$ is regular, hence $B \supset A$ (Theorem 10). The hypothesis is $\lim_A x = L$, $\lim_B x = 0$ so $L = 0$. (A weaker result but with a nice functional analysis proof is given in [80], #2-3-106.)

12. THEOREM. *Let* A, B *be regular and row-finite and assume that* $AB = BA$. *Then* A, B *are consistent.*

Using the associativity result 1.4.5, for $x \in c_A \cap c_B$ we have $\lim_B x = \lim Bx = \lim ABx = \lim BAx = \lim Ax = \lim_A x$.

(f) *MERCERIAN THEOREMS*

13. DEFINITION. *A Mercerian matrix* A *is one such that* $c_A = c$.

The name derives from Mercer's theorem (2.4.1).

14. THEOREM. *A conservative triangle* A *is Mercerian iff* A^{-1} *is conservative.*

By Theorem 5 with $B = I$.

The result fails for non-triangles. (1.8.6).

15. COROLLARY. *Triangles* A, B *are equipotent iff* $B = MA$ *with* M *a Mercerian triangle.*

Sufficiency: $BA^{-1} = M$, $AB^{-1} = M^{-1}$; apply Theorem 5. Necessity: let $M = BA^{-1}$; apply Theorem 5.

It is useful to extend half of this result:

16. THEOREM. *Let* A *be a matrix,* M *a Mercerian triangle,* $B = MA$. *Then* A, B *are equipotent.*

If $x \in c_A$, $Bx = M(Ax) \in c$, using 1.4.4(i). If $x \in c_B$, $(M^{-1}B)x = M^{-1}(Bx) \in c$ using 1.4.4(i). But $M^{-1}B = M^{-1}(MA) = A$ by 1.4.4(i) with A replaced by M and x taken to be an arbitrary column of A. Hence $Ax \in c$.

(g) *MAPPING THEOREMS*

These give characterizations for membership in $(X:Y)$. Examples are $(\ell^\infty:\omega)$, $(c_0:\omega)$, (1.3.2); $(\ell^\infty:\ell^\infty)$, (c_0,ℓ^∞) hence $(c:\ell^\infty)$, (1.3.3) and $(c:c)$, $(c_0:c)$, (1.3.6). A unified treatment using functional analysis is the subject of Chapter 8. We give here a characterization of $(\ell^\infty:c)$, the so-called *coercive matrices*. This mapping theorem is exceptional in that no functional analysis treatment is known. On the contrary an important result (Corollary 20) derives from it. See also [80], Remarks 14-4-8, 15-2-3.

17. LEMMA. *Let* A *be a matrix with convergent columns such that* $\Sigma_k|a_{nk}|$ *is uniformly convergent. Then* $a \in \ell$. (For a see 1.3.5).

This is really just the proof that ℓ is complete. Choose m so that $\Sigma_{k=m}^{\infty}|a_{nk}| < 1$ for all n. Then $\Sigma_k|a_{nk}| < \Sigma_{k=1}^{m-1}|a_{nk}| + 1$ $\to \Sigma_{k=1}^{m-1}|a_k| + 1$ and so $\|A\| < \infty$. The result follows by 1.3.7.

18. THEOREM. *A matrix* A *is coercive iff it has convergent columns and any one of these equivalent conditions holds:*
(i) $\Sigma_k|a_{nk}|$ *is uniformly convergent, (ii) the rows of* A *and* $a \in \ell$, *and* $\Sigma_k|a_{nk}-a_k| \to 0$, *(iii)* $\Sigma_k|a_{nk}| \to \Sigma|a_k|$, *both series being convergent.*

(i) implies (ii): The series is uniformly convergent, so lim and Σ may be interchanged.

(ii) implies (iii): $|\Sigma|a_{nk}|-\Sigma|a_k|| \leq \Sigma|a_{nk}-a_k|$.

(iii) implies (i): Let $\varepsilon > 0$ and $\Sigma_{k=m}^{\infty}|a_k| < \infty$. Choose N so that $n > N$ implies $\Sigma_{k=m}^{\infty}|a_{nk}| < \varepsilon$. This excludes only finitely many n so by increasing m we can ensure that this last inequality holds for all n.

(ii) implies that A is coercive: If x is bounded, $\Sigma_k a_{nk}x_k$ is uniformly convergent so lim and Σ may be interchanged. This yields

$$\lim_A x = ax \quad \text{for bounded } x \tag{1}$$

Next assume that A is coercive. Then it is conservative so the column limits exist (1.3.6) and $a \in \ell$ (1.3.7). We shall prove (iii):

CASE I. *A real,* $a_k = 0$ *for all* k. Suppose (iii) is false. By deleting rows of A and multiplying by a constant we may assume that $\Sigma_k |a_{nk}| > 1$ for all n.

Choose $n(1)$ so that $\Sigma|a_{n(1),k}| > \frac{1}{2}$, $m(1)$ so that $\Sigma_{k=1}^{m(1)} |a_{n(1),k}| > \frac{1}{2}$, $\Sigma_{k=m(1)+1}^{\infty} |a_{n(1),k}| < 1/8$.

Choose $n(2)$ so that $\Sigma_{k=1}^{m(1)} |a_{n(2),k}| < 1/8$, $\Sigma_{k=m(1)+1}^{\infty} |a_{n(2),k}| > 1/2$.

Choose $m(2) > m(1)$ so that $\Sigma_{k=m(1)+1}^{m(2)} |a_{n(2),k}| > 1/2$, $\Sigma_{k=m(2)+1}^{\infty} |a_{n(2),k}| < 1/8$. In general,

choose $n(i)$ so that $\Sigma_{k=1}^{m(i-1)} |a_{n(i),k}| < 1/8$, $\Sigma_{k=m(i-1)+1}^{\infty} |a_{n(i),k}| > 1/2$, and $m(i) > m(i-1)$ so that $\Sigma_{k=m(i-1)+1}^{m(i)} |a_{n(i),k}| > \frac{1}{2}$, $\Sigma_{k=m(i)+1}^{\infty} |a_{n(i),k}| < 1/8$.

Let $x_k = \operatorname{sgn} a_{n(1),k}$ for $1 \le k \le m(1)$; $-\operatorname{sgn} a_{n(2),k}$ for $m(1) < k \le m(2)$; $\operatorname{sgn} a_{n(3),k}$ for $m(2) < k \le m(3)$ and so on.

Then $(Ax)_{n(1)} \ge \Sigma_{k=1}^{m(1)} |a_{n(1),k}| - \Sigma_{k=m(1)+1}^{\infty} |a_{n(1),k}| > 1/2 - 1/8 > 1/4$ (in the second sum $|x_k| = 1$ for all k was used);

$-(Ax)_{n(2)} \ge \Sigma_{k=m(1)+1}^{m(2)} - \Sigma_{k=1}^{m(1)} - \Sigma_{k=m(2)+1}^{\infty} |a_{n(2),k}| > 1/2 - 1/8 - 1/8 = 1/4$ and so on, so that Ax has the subsequence $\{(-1)^n u_n\}$ where $u_n > 1/4$. Thus $x \in \ell^\infty$ yet $Ax \notin c$.

CASE II. *A real.* Let $b_{nk} = a_{nk} - a_k$. Then B is also coercive (1.3.7) and $b_k = 0$ so by case I, B satisfies (iii) which says exactly that A satisfies (ii). This is the same thing, as proved earlier.

CASE III. Let $A = B + iC$. Then for real $x \in \ell^\infty$, $Bx \in c$ since it is the real part of Ax. Hence by Case II, B satisfies the other conditions. Similarly C does and so finally, does A.

We shall see in 5.2.12 that uniform convergence of $\Sigma_k a_{nk}$ is not sufficient.

19. THEOREM. *$A \in (\ell^\infty : c_o)$ iff it has null columns and $\Sigma_k |a_{nk}|$ is uniformly convergent i.e. $(\ell^\infty : c_o) = (\ell^\infty : c) \cap (\phi : c_0)$.*

The other equivalent conditions of Theorem 18 apply here too. Necessity is trivial by Theorem 18 and the fact that $A\delta^k \in c_o$ for each k. Sufficiency is immediate from (1) in the proof of Theorem 18.

20. COROLLARY. *In ℓ, weakly convergent sequences are norm convergent.* Let $\{a^n\}$ be weakly convergent, $A = (a^n_k)$. Then A is coercive since $\ell' = \ell^\infty$ (1.0.2, 1.0.7), and Theorem 18 (ii) says that $a^n \to a$ in norm.

21. COROLLARY. *A coregular matrix cannot sum all the bounded sequences. A coercive matrix must be conull.*

Set $x = 1$ in (1) in the proof of Theorem 18.

1.8 INVERSES

Applications of the inverse of a triangle were given in the preceding section. Some results of this type hold in more generality.

1. LEMMA. *Let $A : c \to c$ be onto. Then $c_A = c \oplus A^\perp$.*

$\supset$: Let $x = u + v$ with $u \in c$, $Av = 0$, which is what the notation means. Then $Ax = Au \in c$. $\subset$: Let $x \in c_A$. Then there exists $u \in c$ with $Au = Ax$. Setting $v = x-u$ gives $x = u+v$ as required.

2. COROLLARY. *If A is conservative and has a conservative right inverse B, $c_A = c \oplus A^\perp$.*

Given $x \in c$ let $u = Bx$. Then $u \in c$ and $Au = (AB)x = x$ using 1.4.4(ii). The result follows from Lemma 1. We shall see in 17.5.11 that the hypothesis of Lemma 1 does not imply that of

Corollary 2.

3. EXAMPLE. Let $(Ax)_n = \Sigma\, 2^{-k} x_{k+2n-2}$, $(Bx)_{2n-1} = 2x_n - x_{n+2}$, $(Bx)_{2n} = -x_{n+1} + 2x_{n+2}$. Then *$A$, B are regular and $AB = I$; A sums the bounded divergent sequence* $\{(-1)^n\}$ as well as unbounded sequences such as $\{n\}$. A matrix A such that $c_A \cap \ell^\infty = c$ is called *Tauberian;* so it appears that the hypothesis of Corollary 2 does not imply that A is Tauberian even if A is regular.

4. THEOREM. *If A is conservative and has a conservative left inverse B, A is Tauberian.*

If $x \in c_A \cap \ell^\infty$, then $x = (BA)x = B(Ax) \in c$ using 1.4.4(ii).

5. EXAMPLE. Let $(Ax)_n = 2x_{n-1} - x_n$, $(Bx)_n = \Sigma\, 2^{-k} x_{k+n}$. *Then A, B are regular, $BA = I$, A is a triangle!, and A sums the divergent sequence* $\{2^n\}$. It appears that the hypothesis of Theorem 4 does not imply that A is Mercerian even if A is a regular triangle. In the next example is shown a non-Mercerian regular matrix with a two-sided regular inverse.

6. EXAMPLE. Let $(Ax)_n = x_n - \alpha x_{n+1}$ where $0 < |\alpha| < 1$. Let $(Bx)_n = \Sigma\, \alpha^{k-1} x_{k+n-1}$. Then $A, B \in \Gamma$, indeed $(1-\alpha)^{-1}A$ and $(1-\alpha)B$ are regular, and $AB = BA = I$. An application of Corollary 2 and Theorem 4 yields the advanced calculus exercise: *suppose $\{x_n - \alpha x_{n+1}\}$ is convergent, $|\alpha| < 1$, and x is bounded; then x is convergent.* Indeed Corollary 2, alone implies that every x such that this transform is convergent is a convergent sequence plus a multiple of $\{\alpha^{-n}\}$. A superficially similar assumption yields a quite different result: *suppose $\{x_n - \alpha x_{n-1}\}$ is convergent, $|\alpha| < 1$; then x is convergent.* This is proved by applying 1.7.14 to the transpose of A. The resulting matrix and its inverse are now triangles.

7. EXAMPLE. *The matrix* A *in Example 5 satisfies* $c_A = c \oplus \{2^n\}$ for if its first row is removed and it is divided by -2 it has the form of Example 6 with $\alpha = \frac{1}{2}$

8. EXAMPLE. *Adjacent convergence domains,* $c_B = c_A \oplus z$. Let A be a triangle and E a row-finite matrix such that $c_E = c \oplus v$ (Example 6, or 7), $B = EA$. If $x \in c_B$, $E(Ax) \in c$ using 1.4.4(i). Thus $Ax = u + mv$ with $u \in c$ and $x = A^{-1}u + mA^{-1}v \in c_A \oplus z$ with $z = A^{-1}v$

A useful sufficient condition for the existence of an inverse is dominance of the principal (or main) diagonal. The idea can be extensively modified, see [85], Theorem 2.

9. THEOREM. *Let* $A \in \Gamma$, $a_{nn} = 1$, $\Sigma\{|a_{nk}| : k \neq n\} < r$ *for all* n *where* $r < 1$. *Then* A *has an inverse in* Γ.

The hypothesis implies that $\|A-I\| \leq r < 1$ and the result follows by 1.0.6, Γ being a Banach algebra (1.5.4).

10. COROLLARY. *Let* A *be a conservative triangle with* $a_{nn} = 1$, $\|A\| < \Sigma\, |a_k| + 2$. *Then* A *is Mercerian.*

Let $\Sigma_{k=1}^{u} |a_k| > \alpha > \|A\| - 2$ and choose v so that $n > v$ implies $\Sigma_{k=1}^{u} |a_{nk}| > \alpha$. Let $m = \max(u,v)$. Now let $b_{nn} = 1$, $b_{nk} = 0$ for all n, k such that $n > m$ and $k \leq m$, $b_{nk} = a_{nk}$ otherwise. Then $c_B = c_A$ and the result follows from Theorem 9 when we show that $\|B-I\| < 1$. To see this: $\Sigma_{k=1}^{n-1} |b_{nk}|$ $= \Sigma_{k=m+1}^{n-1} |a_{nk}| = \Sigma_{k=1}^{n} |a_{nk}| - 1 - \Sigma_{k=1}^{m} |a_{nk}| < \|A\| - 1 - \alpha < 1.$

11. EXAMPLE. For $A = (C,1)$ the second condition holds but $a_{nn} \to 0$; for $A = 2Q$ (1.2.5), $a_{nn} = 1$ but $\|A\| = \Sigma\, |a_k| + 2$; for $(Ax)_n = x_n - \frac{1}{2} x_{n-1}$, Corollary 10 applies.

The next example is an important application:

12. EXAMPLE (S. MAZUR'S MATRIX). Let $t \in \ell$ and let $M = M(t)$ be the matrix with $(Mx)_n = x_n + \Sigma_{k=1}^{n-1} t_k x_k$; $\|M\| = 1 + \Sigma|t_k|$ and $m_k = t_k$ so Corollary 10 implies that *M is Mercerian.* Note also that *for $x \in c$, $\lim_M x = \lim x + tx$ (1.3.8).*

CHAPTER 2
CLASSICAL MATRICES

2.0. BACKGROUND

This section is exceptional, as is the whole chapter, in that it is classical. See §2.1.

1. (Helly's choice theorem). Let g_n be an increasing function on $[0,1]$ for each n and assume that $|g_n(t)| < M$ for each n, t. Then $\{g_n\}$ has a pointwise convergent subsequence. [74], 9.6I

2. THEOREM. With g_n as in Theorem 1 suppose that $g_n \to g$ pointwise. Then $\int f dg_n \to \int f dg$ for each continuous f. [74], 9.6II

3. (G.H. Hardy's big O Tauberian theorem). Let y be (C,1) summable and such that $\{n(y_n - y_{n-1})\}$ is bounded. Then y is convergent. [90], p. 225.

The reason for the name is the notation $x_n = O(t_n)$ to mean $\{x_n/t_n\}$ is bounded.

2.1 INTRODUCTION

The reader who is interested mainly in functional analysis may proceed immediately to Chapter 3. The rest of the book makes few references to the present chapter and these can easily be consulted as needed.

The classical matrices were introduced in the 19th century for application to problems in analysis such as analytic continuation of power series and improvement of the rate of convergence of numerical series.

2.2 HÖLDER MATRICES

Let $H = (C,1)$. The Hölder matrix of order n, $n = 1,2,\ldots$, is simply H^n, the nth power of H. Since $H^{n+1}(H^n)^{-1} = H$ it follows from 1.7.5 that $H^m \supset H^n$ if $m > n$. The inclusion is strict, by 1.75 ($A = H^2$, $B = H$) and 1.7.14 ($A = H$) since H sums the divergent sequence $\{(-1)^n\}$; it is also obvious from 1.7.8. We have, however, the following Tauberian theorem:

1. THEOREM. *All* H^n, $n = 1,2,\ldots$, *are equivalent on* ℓ^∞.

This means that if x is bounded and summable H^m it is summable H^n to the same value for all n. Suppose first that x is H^2 summable and let $y = Hx$. Then $n(y_n - y_{n-1}) = x_n - y_{n-1}$ which is bounded hence $y \in c$, (2.0.3), i.e. $x \in c_H$. If x is H^{k+1} summable, $k > 1$, let $y = H^{k-1}x$. Then y is H^2 summable, hence H summable as just proved and so x is H^k summable. The limits are equal as just mentioned.

2.3. HAUSDORFF MATRICES

All indices will start from 0 *in this section.* Let $\mu = (\mu_0, \mu_1, \mu_2, \ldots)$ be a complex sequence. Let M be the diagonal matrix with $m_{nn} = \mu_n$ for $n = 0,1,2,\ldots$. Let D be the matrix $d_{nk} = (-1)^k\binom{n}{k}$ (binomial coefficients.) Then D is a triangle with $d_{nn} = (-1)^n$, $d_{no} = 1$ etc. A little calculation will suggest that $D^2 = I$. We shall prove this shortly.

The matrix $H_\mu = DMD$ is called the *Hausdorff matrix associated with the sequence* μ. It is triangular since D and M are; and it is a triangle iff $\mu_n \neq 0$ for all n since $h_{nn} = \mu_n$. (When μ is fixed throughout a discussion we shall write H for H_μ.)

To facilitate computation in this section we introduce two matrices. Let S be the matrix such that $(Sx)_n = x_{n+1}$ for all n, so, for example, the first two rows of S are $(0,1,0,0,\ldots)$

and $(0,0,1,0,0,\ldots)$. (Of course $x = (x_0,x_1,x_2,\ldots)$.) We shall use the formula $(S^i x)_k = x_{i+k}$. Let $\Delta = I - S$ so that $(\Delta x)_n = x_n - x_{n+1}$.

By the definition $h_{nk} = 0$ if $h > n$; and for $k \le n$:

$$h_{nk} = \sum_{j=k}^{n} (-1)^{j+k}\binom{n}{j}\binom{j}{k}\mu_j = \binom{n}{k}\sum_{j=k}^{n}(-1)^{j+k}\binom{n-k}{j-k}\mu_j$$

$$= \binom{n}{k}\sum_{i=0}^{n-k}(-1)^i\binom{n-k}{i}\mu_{i+k} \qquad (1)$$

The last step was arrived at by the substitution $i = j-k$, noting that $(-1)^{i+2k} = (-1)^i$. Continuing,

$$h_{nk} = \binom{n}{k}\sum_{i=0}^{n-k}(-1)^i\binom{n-k}{i}(S^i\mu)_k = \binom{n}{k}(\Delta^{n-k}\mu)_k \qquad (2)$$

There are no symbolic operations here; S^i and Δ^{n-k} are simply powers of matrices; the matrices are row finite so all multiplication is associative.

1. EXAMPLE. Let $\mu = 1$. Then $\Delta^0\mu = \mu = 1$, and, for $r > 0$, $\Delta^r\mu = 0$. Thus $h_{nn} = 1$, $h_{nk} = 0$ if $k \neq n$ i.e. $H_1 = I$. Thus, as promised, $D^2 = D1D = H_1 = I$.

2. THEOREM. $H_\mu H_\nu = H_{\mu\nu}$.

For the left side is DMDDND = DMND.

3. COROLLARY. *$(H_\mu)^r = H_{\mu^r}$ for $r = 0,1,2,\ldots$. If $\mu_n \neq 0$ for all n, $(H_\mu)^{-1} = H_{(1/\mu)}$.*

4. THEOREM. *All Hausdorff matrices commute with each other. Hence all regular Hausdorff matrices are consistent.*

This follows from Theorem 2 and 1.7.12.

5. EXAMPLE. Fix $t \in K$ and set $\mu_n = t^n$. Then $\Delta\mu = (1-t)\mu$ so $\Delta^r\mu = (1-t)^r\mu$ and $h_{nk} = \binom{n}{k}(1-t)^{n-k}t^k$ from (2).

6. EXAMPLE. Let $\mu_n = 1/(n+1)$. Then $\mu_n = \int_0^1 t^n dt$ so $h_{nk} = \int_0^1 h_{nk}(t)dt$ (where $h_{nk}(t)$ is as in Example 5)

$= \binom{n}{k} \int_0^1 (1-t)^{n-k} t^k dt = 1/(n+1)$. Thus $H_\mu = (C,1)$. So *the Cesàro matrix is a Hausdorff matrix.* Note that to agree with the convention of this section, subscripts start from 0 and the Cesàro matrix is altered accordingly.

7. EXAMPLE. The matrix Q (1.2.5) is not a Hausdorff matrix by Theorem 2 since it does not commute with (C,1).

8. EXAMPLE. *The Hölder matrices are all Hausdorff matrices.* Indeed $H^r = H_\mu$ with $\mu_n = (n+1)^{-r}$. This follows from Theorem 2 and Example 6.

9. THEOREM. *Let* μ *be a one-to-one sequence (i.e.* $\mu_m \neq \mu_n$*) and* A *a row-finite matrix. Then* A *is a Hausdorff matrix iff* A *commutes with* H_μ.

Half of this is Theorem 4. If A commutes with H we have DAHD = DHAD. Substituting DMD for H and using $D^2 = I$ we have DADM = MDAD. It is obvious that only a diagonal matrix can commute with M so DAD is a diagonal matrix; call it N. Then using $D^2 = I$ we have A = DND, a Hausdorff matrix.

W.A. Hurwitz and L.L. Silverman (1917), wishing to consider only summability methods consistent with (C,1), applied the commutativity criterion, 1.7.12, and investigated *a priori* those matrices commuting with (C,1), a special case of Theorem 9. In 1921 F. Hausdorff introduced his matrices from the point of view of an analyst interested in the moment problem, as described below.

10. LEMMA. *Let* $H = H_\mu$. *Then for all* $n \geq 0$ *(i)* $h_{nn} = \mu_n$, *(ii)* $\sum_{k=0}^{n} h_{nk} = \mu_0$, *(iii)* $h_{no} = (D\mu)_n$, *(iv)* $\sum_{k=0}^{m} h_{nk} - \sum_{k=0}^{m} h_{n+1,k}$ $= \frac{m+1}{n+1} h_{n+1,m+1}$ *for all* $m \geq 0$.

For (ii), $H1 = DMD1 = DM\delta^o = D(\mu_o\delta^o) = \mu_o(D\delta^o) = \mu_o 1$. Here we used the facts that 1 is the first column of D and $D^2 = I$. Setting $k = 0$ in (1) yields (iii). Considering (iv), keep in mind that $h_{nk} = 0$ where $k > n$. We have from (2)

$$\frac{h_{n+1,k}}{\binom{n+1}{k}} = (\Delta^{n-k+1}\mu)_k = (\Delta\Delta^{n-k}\mu)_k = (\Delta^{n-k}\mu)_k - (\Delta^{n-k}\mu)_{k+1}$$

$$= \frac{h_{nk}}{\binom{n}{k}} - \frac{h_{n+1,k+1}}{\binom{n+1}{k+1}}$$

Multiply this by $\binom{n}{k}$; subtract $h_{n+1,k}$ and transpose: $h_{nk} - h_{n+1,k} = [(k+1)h_{n+1,k+1} - kh_{n+1,k}]/(n+1)$. This yields (iv).

11. THEOREM. *Every non-negative (real) Hausdorff matrix* H *is conservative. All of its column limits are* 0 *except possibly the first.*

From Lemma 10 (ii) it follows that conditions (i) and (iii) of Theorem 1.3.6 hold. Induction on m in Lemma 10 (iv) yields the remaining condition. (Note that $\sum_{k=0}^{m} h_{nk}$ is a decreasing non-negative function of n since $h_{n+1,m+1} \geq 0$.) Next, Lemma 10 (iv) yields, for any r, (write $u_n = h_{n+1,m+1}$, holding m fixed),

$$\sum_{n=0}^{r} u_n/(n+1) = [\sum_{k=0}^{m} (h_{ok}-h_{r+1,k})]/(m+1).$$

The right hand side converges as $r \to \infty$, as just proved, so $\Sigma\, u_n/(n+1) < \infty$. Since $\lim u_n$ exists it must be 0.

12. EXAMPLE. Let $\mu = \delta^o = (1,0,0,\ldots)$. Then H is the matrix whose first column is 1 and all of whose other columns are 0.

13. DEFINITION. A sequence μ is called *totally decreasing* if H_μ is a non-negative matrix.

By (2) the condition is $(\Delta^n)\mu \geq 0$ for all n. For $n = 0$ it says $\mu \geq 0$; for $n = 1$ it says $\mu_0 \geq \mu_1 \geq \mu_2 \geq \ldots$, for $n = 2$ it says $\mu_0 - 2\mu_1 + \mu_2 \geq 0$, $\mu_1 - 2\mu_2 + \mu_3 \geq 0$, etc., a convexity condition. The sequence $(3,2,0,0,0,\ldots)$ is not totally decreasing. The sequences in Example 5 $(0 < t < 1)$, 6, 8 are totally decreasing.

14. THEOREM. *Equivalent conditions for a real Hausdorff matrix* $H = H_\mu$ *are (i)* H *is conservative, (ii)* $\|H\| < \infty$*, (iii)* H *is the difference of two non-negative Hausdorff matrices, (iv)* μ *is the difference of two totally decreasing sequences.*

It is sufficient to prove that (ii) implies (iv). Now (iv) will follow from the existence of a sequence ν such that

$$\Delta^n \nu > |\Delta^n \mu| \tag{3}$$

since with $\alpha = \frac{1}{2}(\nu+\mu)$, $\beta = \frac{1}{2}(\nu-\mu)$ we have $\mu = \alpha - \beta$. We turn to the construction of ν. The idea is a simple one. Consider the matrix D given by $d_{nk} = (\Delta^n\mu)_k$. We express μ as a function of D and simply define ν as the same function of $|D|$.

As in the proof of Lemma 10 (iv) we have $d_{nk} = d_{n+1,k} + d_{n,k+1}$ i.e. $D = UD + TD = (U+T)D$ where U, T are the (symbolic) operators $(UD)_{nk} = d_{n+1,k}$, $(TD)_{nk} = D_{n,k+1}$. So, for any m, $D = (U+T)^m D = \sum_{r=0}^{m} \binom{m}{r} U^{m-r} T^r D$, i.e. $d_{nk} = \sum_{r=0}^{m} \binom{m}{r} d_{n+m-r,k+r}$. It follows that

$$\mu_k = (\Delta^0 \mu)_k = d_{ok} = \sum_{r=0}^{m} \binom{m}{r} d_{m-r,k+r} \tag{4}$$

Now, as predicted, we set $A = |D|$ and use it to define ν.

Let $f(m,n,k) = \sum_{r=0}^{m} \binom{m}{r} a_{n+m-r,k+r} = (U+T)^m A$. We shall prove in Lemma 15 that $f(m,n,k)$ is an increasing bounded function of m for each n, k. Let $g(n,k) = \lim f(m,n,k)$ and define $\nu_k = g(0,k)$. The proof of (3) is: $|(\Delta^n \mu)_k| = |d_{nk}| = |\sum \binom{m}{r} d_{n+m-r,k+r}|$ $\leq f(m,n,k) \leq g(n,k) = (\Delta^n \nu)_k$. (The last step is proved in Lemma 16)

15. LEMMA. *$f(m,n,k)$ is an increasing bounded function of m for each n, k.*

Note that $f = [(U+T)^m A]_{nk}$; $A = |D| = |UD+TD| \leq UA + TA$ $= (U+T)A$. The operator $U + T$ is monotone, so $(U+T)A \leq$ $(U+T)(U+T)A = (U+T)^2 A$, i.e. $f(1,n,k) \leq f(2,n,k)$ and so on. Next we observe that $f(m,n,k) = [(U+T)^m U^n T^k A]_{oo} \leq [(U+T)^m (U+T)^{n+k} A]_{oo}$ $= f(m+n+k,0,0)$. The inequality holds because $[(U+T)^{n+k} A]_{oo}$ $= \sum_{r=0}^{n+k} \binom{n+k}{r} a_{r,n+k-r} \geq \binom{n+k}{n} a_{n,k} \geq a_{nk} = (U^n T^k A)_{oo}$. Finally $f(m,0,0) = \sum_{r=0}^{m} |h_{mr}| \leq \|H\|$.

16. LEMMA. *$g(n,k) = (\Delta^n \nu)_k$.*

By induction on n (check $n = 0$ against the definition): first, $f(m+1,n,k) = [(U+T)^m (U+T)A]_{nk} = [(U+T)^m A]_{n+1,k}$ $+ [(U+T)^m A]_{n,k+1} = f(m,n+1,k) + f(m,n,k+1)$ and so $g(n,k)$ $= g(n+1,k) + g(n,k+1)$. Hence, using the induction hypothesis, $g(n+1,k) = g(n,k) - g(n,k+1) = (\Delta^n \nu)_k - (\Delta^n \nu)_{k+1} = (\Delta^{n+1} \nu)_k$.

The next definition will not be used outside this chapter.

17. DEFINITION. *Call μ conservative if H_μ is conservative, regular if H_μ is regular.*

Thus a real sequence is conservative iff it is the difference of two totally decreasing sequences. We proceed to the solution of the *moment problem*.

18. THEOREM. *A real sequence* μ *is conservative iff there exists a function* g *of bounded variation on* $[0,1]$ *such that* $\mu_k = \int_0^1 t^k dg$.

Sufficiency: We may assume that g is increasing. Then, as in Example 5,

$$(\Delta^n\mu)_k = \int_0^1 t^k(1-t)^n dg > 0. \tag{5}$$

Necessity: We may assume that μ is totally decreasing. By (4), $\mu_k = \sum_{r=0}^{m} \binom{m}{r} h_{m+k,k+r}/\binom{m+k}{k+r}$; replacing m by $m-k$, we have, for $m \geq k$, $\mu_k = \sum_{i=k}^{m} \binom{m-k}{i-k} h_{mi}/\binom{m}{i} = \sum_{i=k}^{m} u(m,i,k)h_{mi}$ where $u(m,i,k) = \prod_{r=0}^{k-1} (i-r)/(m-r)$ for $k \leq i \leq m$, with the convention $u(m,i,0) = 1$. We may write $\mu_k = \sum_{i=0}^{m} u(m,i,k)h_{mi}$ because $u = 0$ if $0 \leq i < k$. Then $\mu_k = \int_0^1 v(m,t,k)dg_m$ where $v(m,t,k) = \prod_{r=0}^{k-1} (t-r/m)/(1-r/m)$, $v(m,t,0) = 1$, and $g_m(0) = 0$, $g_m(t) = \sum_{j\leq mt} h_{mj}$ for $0 < t \leq 1$. This is because g_m has jump h_{mi} at $t = i/m$ for $i = 0,1,\ldots$. Note also that g_m is increasing. For each fixed $h \geq 1$, $|(t-r/m)/(1-r/m)-t| = r(1-t)/(m-r) \leq (k-1)/(m-k+1)$ since $r < k \leq m$, $0 \leq t \leq 1$; thus $v(m,t,k) \to t^k$ uniformly. Since also $\int dg_m = g_m(1) - g_m(0) = \sum_i h_{mi} \leq \|H\|$ we have $\mu_k = \int_0^1 t^k dg_m + o(1)$ as $n \to \infty$ where $o(1)$ denotes a quantity which $\to 0$ as $m \to \infty$.

In this formula we may let $m \to \infty$ through a sequence of values such that $g_m \to g$ (some increasing function). This is by 2.0.1. The result is $\mu_k = \int_0^1 t^k dg$.

19. THEOREM. *Let* μ *be conservative,* $\mu_k = \int_0^1 t^k dg$. *The first column limit of* H_μ *is* $g(0+)-g(0)$. *Thus* H *is multiplicative*

(with $m = g(1)-g(0)$) iff g is continuous at 0.

We may assume that g is increasing and that $g(0) = 0$. By (5), $h_{no} = \int_0^1 (1-t)^n dg \geq \int_0^\varepsilon (1-t)^n dg \geq (1-\varepsilon)^n g(\varepsilon) \to g(0+)$ as $\varepsilon \to 0$. Conversely $h_{no} = \int_0^\varepsilon + \int_\varepsilon^1 (1-t)^n dg \leq g(\varepsilon) + (1-\varepsilon)^n[g(1)-g(\varepsilon)] \to g(\varepsilon)$ as $n \to \infty$. Thus $\lim h_{no} \leq g(\varepsilon)$ for all $\varepsilon > 0$. The evaluation of m follows from Lemma 10 (ii).

20. EXAMPLE. (i) Taking $g = 0$ on $[0,1)$, $g(1) = 1$ yields $H = I$. (ii). With $g(t) = t$ we obtain $H = (C,1)$.

We now show Hausdorff's elegant derivation of Mercerian theorems. First with $g(t) = t^a$, $a > 0$, $\mu_k = \int_0^1 t^k dg = a/(k+a)$. Then μ is regular by Theorems 18, 19, so for any $\alpha > 0$, $\nu = \alpha 1 + (1-\alpha)\mu$ is regular since $H_\nu = \alpha I + (1-\alpha)H_\mu$ by Example 1. Note that $\nu_k = (\alpha k+a)/(k+a)$.

21. THEOREM. *Let $\mu_k = (bk+1)/(ck+1)$, $b,c > 0$. Then H_μ is regular and Mercerian.*

The substitution $b = \alpha/a$, $c = 1/a$ makes $\mu_k = (\alpha k+a)/(k+a)$ which, as we just saw, is regular. Similarly, $1/\mu$ is regular and the result follows by Corollary 3 and 1.7.14.

22. COROLLARY. *Let $\mu(a)_k = (ak+1)^{-1}$. Then $H_{\mu(a)}$ and $H_{\mu(b)}$ are equivalent for all $a,b > 0$.*

By 1.7.10.

A special case of Hausdorff's theorem is Mercer's theorem which we give in the next section.

Hausdorff matrices are discussed also in [55], pp. 167-173.

2.4. CESÀRO AND HÖLDER

Mercer's theorem (1907) states that the straight line joining the identity matrix to $(C,1)$ is made up of Mercerian matrices. We shall write C for $(C,1)$ and remind the reader that indices start from 0.

1. THEOREM. *For* $\alpha > 0$, *the matrix* $\alpha I + (1-\alpha)C$ *is Mercerian.*

For it is H_μ with $\mu_k = \alpha + (1-\alpha)/(k+1) = (\alpha k+1)/k+1$. The result follows by 2.3.21.

For $\alpha = 0$ the matrix, call it A, in Theorem 1 is $(C,1)$; for $\alpha < 0$ A is not Mercerian, but has a very small convergence domain:

2. THEOREM (G.H. Hardy, 1913). *Let* $\alpha < 0$, $A = \alpha I + (1-\alpha)C$, $X = c_A$. *Then* $X = c \oplus v$ *where* v *is a divergent sequence.*

The notation means that every A-summable sequence is a convergent sequence + a multiple of v and conversely. (To motivate the choices about to be made, let q be the first column of CA^{-1}. Then $AC^{-1}q = \delta^o$, hence, since A, C commute, $Aq = C\delta^o$ so $q \in X$. Computation of q — which will be a little easier using $(AC^{-1})^{-1}$ — will reveal the origin of the sequence v defined in the following argument.)

Let $x \in X$. Since $x = u+L.1$ where $L = \lim_A x$ and $u = x-L.1$ satisfies $\lim_A u = 0$ we may assume that $\lim_A x = 0$. Let $\lambda = 1 - \alpha^{-1} > 1$; $y_n = (n+1)(Cx)_n = \sum_{i=0}^{n} x_i$; $z_n = y_n\Gamma(n+2-\lambda)/\Gamma(n+2)$ for all n if λ is not an integer, while if λ is an integer, define z_n this way for $n \geq \lambda-1$; for smaller n, z_n may be defined arbitrarily. For sufficiently large n,

$$\begin{aligned} z_n-z_{n-1} &= [y_n(n+1-\lambda)/(n+1)-y_{n-1}].\Gamma(n+1-\lambda)/\Gamma(n+1) \\ &= \alpha^{-1}(Ax)_n\Gamma(n+1-\lambda)/\Gamma(n+1) = o(n^{-\lambda}). \end{aligned}$$

Thus $\sum (z_n - z_{n-1})$ is convergent so that $M = \lim z_n$ exists and $z_n = M - \sum_{i=n+1}^{\infty} (z_i - z_{i-1}) = M + o(n^{1-\lambda})$. Hence $x_n = \alpha^{-1}(Ax)_n + \lambda y_n/(n+1) = o(1) + \lambda z_n \Gamma(n+1)/\Gamma(n+2-\lambda) = o(1) + \lambda z_n v_n$, say. This is $o(1) + \lambda M v_n + \lambda v_n . o(n^{1-\lambda}) = \lambda M v_n + o(1)$. This proves that $X \subset c \oplus v$ and the reverse inclusion merely requires $v \in X$. To see this: for large n, $(C^{-1}v)_n = (n+1)v_n - nv_{n-1} = \lambda v_n$, so $(Av)_n = \alpha v_n + (1-\alpha)\lambda^{-1} v_n = 0$.

Given a matrix A, let $M(A)$ (the Mercerian set) be $\{\lambda : \lambda I + (1-\lambda)A$ is Mercerian$\}$. In [45] it is shown that there is a regular triangular A such that $M(A)$ is any preassigned open set containing 1 — but there is also such an A that $M(A)$ is not open.

3. DEFINITION. *The Cesàro matrix of order* α, *written* C_α, *for* $\alpha > 0$, *is the Hausdorff matrix* H_μ in which $\mu_k = \alpha \int_0^1 t^k (1-t)^{\alpha-1} dt$.

(The definition of C_α is usually extended to a larger class of values of α.) By 2.3.18 with $g(t) = -(1-t)^\alpha$, each C_α is regular.

Setting $H = C_\alpha$ a straightforward calculation gives $h_{nk} = \binom{n-k+\alpha-1}{n-k} / \binom{n+\alpha}{n}$, $h_{kk} = \mu_k = \binom{k+\alpha}{k}^{-1}$. It is natural to compare this with the Hölder matrix H^α which has $\mu_k = (k+1)^{-\alpha}$, $H^1 = C_1 = (C,1)$. Also if A is the matrix such that $(Ax)_n = \sum_{k=0}^{n} x_k$ and $B = \operatorname{diag}(1/n)$ we have, for integers α, $H^\alpha = (BA)^\alpha$, $C_\alpha = B^\alpha A^\alpha$. A more intimate connection is the theorem of Schnee and Knopp ([34], p. 264) which states that these matrices are equivalent. We give it for integer α only.

4. THEOREM. *The matrices* H^α, C_α *are equivalent for* $\alpha = 1,2,\ldots$.

They are H_μ, H_ν with $\mu_k = (k+1)^{-\alpha}$, $\nu_k = \binom{k+\alpha}{k}^{-1}$. Now $\mu_k/\nu_k = (\alpha!)^{-1}\ \pi\{(k+r)/(k+1):r = 2,3,\ldots,\alpha\}$, thus $H^\alpha(C_\alpha)^{-1}$ is a product of matrices of the form H_λ with $\lambda = (k+r)/(k+1)$. By 2.3.21 each such matrix is Mercerian.

Some discussion and references for complete monotonicity may be found in Problem E2845 of the American Mathematical Monthly, January 1982, p. 64.

2.5. NÖRLUND MATRICES

We introduce the notation (a,b) for the sequence formed from two sequences a,b by the formula $(a,b)_n = \sum_{k=0}^{n} a_k b_{n-k}$. (This is the standard *convolution.*) *Indices start from* 0 *everywhere in this chapter.*

Let p be a complex sequence with $p_o = 1$. Let $P_n = (p,1)_n = \sum_{k=0}^{n} p_k$. We shall assume that $P_n \neq 0$ for all n. The *Nörlund* matrix $A = (N,p)$ is defined by the formula $a_{nk} = p_{n-k}/P_n$ for $k \leq n$, 0 for $k > n$. Thus $(Ax)_n = (p,x)_n/(p,1)_n$, $\sum_k a_{nk} = (A1)_n = 1$.

Manipulations with Nörlund matrices are conveniently handled via the formal power series $p(z) = \sum_{n=0}^{\infty} p_n z^n = (1-z)P(z)$ where $P(z) = \sum P_n z^n$. For obvious reasons (N,p) is called a *polynomial matrix* if $p_n = 0$ for sufficiently large n. The functions $p(z)$ may be defined first and the Nörlund matrix defined from it. This is done in Example 1. It is important to remember that whenever $p(z)$ is given and generates a Nörlund matrix (N,p), parts of the data are $p(0) = p_o = 1$, $P_n \neq 0$.

1. EXAMPLE. (a) Let $p(z) = 1$. Then $p = \delta^o$, $(N,p) = I$. (b) Let $p(z) = 1+z$. Then $(N,p) = (N,1+z) = Q(1.2.5)$, (c) Let $p(z) = (1-z)^{-1}$. Then $p = 1, (N,p) = (C,1)$.

2. EXAMPLE. For positive integer α, let $p(z) = (1-z)^{-\alpha} = \sum \binom{n+\alpha-1}{n} z^n$. Then $P(z) = p(z)/(1-z) = (1-z)^{-\alpha-1} = \sum \binom{n+\alpha}{n} z^n$ and $(N,p) = C_\alpha$. The latter are the only matrices which are both Nörlund and Hausdorff!

3. THEOREM. *Let A be a Nörlund and a Hausdorff matrix. Then $A = C_\alpha$ for some α.* Let $A = (N,p) = (H,\mu)$. Then, setting $\alpha = p_1$, $\alpha\mu_n = \alpha a_{nn} = p_1/P_n = a_{n,n-1} = n(\mu_{n-1}-\mu_n)$. Thus $\mu_n = (n/(n+\alpha))\mu_{n-1}$. Since $\mu_o = a_{oo} = 1$ it follows by induction that $\mu_n = \binom{n+\alpha}{n}^{-1}$.

4. Let $A = (N,p)$ have finite norm. Then for each n, $\|A\| > |a_{nn}| = |P_n^{-1}|$. Thus P_n is bounded away from 0. It follows that $p(1) \neq 0$. (In Example 1(c), $p(1)$ is not defined.) For a polynomial matrix, $p(1) = P_u$ where u is the degree of p. This was explicitly assumed non-zero. Note again that $p(0) = p_o = 1$.

5. THEOREM. *Every polynomial matrix is regular.*

The matrix consists of finitely many diagonals. Each column terminates in 0's; each row adds up to 1. Finally, for sufficiently large n, $\sum_k |a_{nk}| = \sum_{k=0}^{m} |p_k/P_m|$ where $p_k = 0$ for $k > m$. Thus $\|A\| < \infty$.

6. THEOREM. *The Nörlund matrix $A = (N,p)$ is conservative iff (i) $\lim p_n/P_n$ exists (call it λ), (ii) $\sum_{k=0}^{n} |p_k| \leq M|P_n|$ for each n; regular iff (i), (ii) hold with $\lambda = 0$.*

Note that

$$a_{no} = p_n/P_n = 1 - P_{n-1}/P_n \tag{1}$$

If A is conservative, (i) follows by (1). Condition (ii) is $\|A\| < \infty$. It remains to prove that (i) implies that $a_k = \lim_n a_{nk}$ exists for each k. Reasoning by induction we have $a_{n,k+1} = p_{n-k-1}/P_n = (p_{n-k-1}/P_{n-1})(P_{n-1}/P_n) = (a_{n-1,k})(P_{n-1}/P_n) \to a_k(1-\lambda)$ as $n \to \infty$. This completes the proof that A is conservative. We also have $a_{k+1} = a_k(1-\lambda)$ so $a_k = a_o(1-\lambda)^k = \lambda(1-\lambda)^k$ by (1); so A is regular if and only if $\lambda = 0$.

7. By (1), condition (i) of Theorem 6 can be replaced by: $\lim P_{n-1}/P_n$ exists. (It is $1-\lambda$). Also by 1.3.7, the conditions of Theorem 6 imply that $\lambda = 0$ or $|1-\lambda| < 1$. Finally, if $\lambda \neq 0$, the proof of Theorem 6 shows that $\Sigma\, a_k = 1 = \lim_A 1$; hence *a conservative Nörlund matrix must be conull or regular.*

8. COROLLARY. *Let $p_n \geq 0$ for all n. Then (N,p) is conservative if and only if $\lim(p_n/P_n)$ exists; regular if and only if $p_n/P_n \to 0$*

9. COROLLARY. *Each of the following is sufficient that (N,p) be regular: (i) $p_n \geq 0$, $\{p_n\}$ bounded; (ii) $p_n \geq 0$, $p_n \to 0$; (iii) $\Sigma\, |p_n| < \infty$, $\Sigma\, p_n \neq 0$; (iv) $p \in \phi$.*

Sufficiency of (i) and (iii) follow from Corollary 8 and Theorem 6, respectively, while (ii) is a special case of (i). The last is just Theorem 5.

We now relate properties of (N,p) with those of $p(z)$, $P(z)$.

10. THEOREM. *Let (N,p) be conservative and R the radius of convergence of $P(z)$. Then $R \leq 1$. Further (N,p) is regular if and only if $R = 1$.*

Using Theorem 6 and its proof we have $R = \lim |P_{n-1}/P_n| = |1-\lambda| \leq 1$ by Remark 7. Also $R = 1$ if and only if $\lambda = 0$.

But $P(z) = (1+z)^{-1}$ has $R = 1$ and (N,p) is not even conservative.

11. THEOREM. *Given $p(z)$, $q(z)$ generating (N,p) and (N,q), let $r(z) = p(z)q(z)$. Then r, the sequence generating the Nörlund matrix (N,r) is given by $r_n = (p,q)_n$, $R_n = (r,1)_n = (p,Q)_n = (P,q)_n$.*

This is standard multiplication of power series.

2.6. POLYNOMIAL MATRICES

1. THEOREM. *Let p, q be polynomials generating Nörlund matrices $A = (N,p)$, $B = (N,q)$, and let $r(z) = p(z)q(z)$. Then $(N,r) = AB + F$ where F is a finite matrix ie. there exists d such that $(N,r)_{n,k} = (AB)_{nk}$ for $n > d$. In particular $c_{AB} = c_{(N,r)}$.*

Of course, $(N,r)_{nk} = (AB)_{nk}$ for $k > d$ (when the theorem has been proved.)

Let $d = u + v$ where u, v are the degrees of p,q. For $n \geq d$ and $k \leq n$, $(AB)_{nk} = \sum_{m=k}^{n} a_{nm}b_{mk} = \sum_{m=k}^{n} (p_{n-m}q_{m-k})/P_nQ_m$ $= (p,q)_{n-k}/P_uQ_v$ because $n \geq u$ so $P_n = P_u$, and, in the range of summation, either $m \geq v$ so $Q_m = Q_v$ or else $m < v$ so $p_{n-m} = 0$ since $n-m > u$.

We show that the same is true for $C = (N,r)$. By 2.5.11 it is sufficient to note that for $n \geq d$, $R_n = (p,Q)_n = \sum_{i=0}^{n} p_iQ_{n-i}$ $= \sum_{i=0}^{u} p_iQ_{n-i} = Q_vP_u$ since $n-i \geq n-u = v$.

2. COROLLARY. *All polynomial Nörlund matrices are consistent.*

It follows from Theorem 1 that $AB-BA$ is a finite matrix, hence sums all sequences to 0. The result follows as in 1.7.12.

We now give the *Kubota-Petersen* theorem which characterizes c_A if $A = (N,p)$ and $p(z)$ is a polynomial with no zeros on $(|z| = 1)$. The idea is that $p(z) = C.\pi(z-\alpha_i)$ and Theorem 1 allows us to treat the factors separately.

3. LEMMA. *Let* $p(z) = 1 - tz$ *with* $|t| < 1$. *Then* (N,p) *is Mercerian.* This is the second part of 1.8.6.

4. LEMMA. *Let* $A = (N,p)$ *be a polynomial matrix and suppose that* $p(\alpha) = 0$. *Then* $x = \{\alpha^{-n}\}$ *is summable* A.

Note by 2.5.4 that $\alpha \neq 0,1$. Now $(Ax)_n = (p,x)_n/P_n$ and $(p,x)_n = \Sigma\, p_i x_{n-i} = \alpha^{-n}p(\alpha) = 0$ for sufficiently large n.

5. THEOREM. *Let* $A = (N,p)$ *be a polynomial matrix. Then* A *is Mercerian if and only if* p *has no zeros in* $(|z| \leq 1)$.

Sufficiency is by Theorem 1 and Lemma 3 together with the obvious fact that a finite product of Mercerian triangles is Mercerian. Necessity is by Lemma 4 ($\alpha \neq 0,1$).

6. LEMMA. *Let* $p(z) = 1 - tz$ *with* $|t| > 1$ *and let* $A = (N,p)$. *Then* $c_A = c \oplus \{t^n\}$.

If the first row of A is dropped and the resulting matrix is multiplied by $(t-1)/t$, the first matrix of 1.8.6 appears with $\alpha = 1/t$.

7. LEMMA. *Let* B *be a triangle and* A *as in Lemma 6,* $D = AB$. *Then* $c_D = c_B \oplus y$ *where* $y = B^{-1}(\{t^n\})$.

First $c_D \supset c_B$ since A is conservative. Also $y \in c_D$ since $By \in c_A$. Conversely if $x \in c_D$ then $Bx \in c_A$ so by Lemma 6,

$Bx = u + v$, where $u \in c$ and v is a multiple of $\{t^n\}$, so $x = B^{-1}u + B^{-1}v$.

Now in order to handle (N,p) we can factor $p(z)$ into linear factors and write $(N,p) = ABC\ldots$ modulo a finite matrix as in Theorem 1 where each of A,B,C is a first degree polynomial matrix. We begin with quadratic $p(z)$ and there are a few cases; $p(z) = (1-sz)(1-tz)$:

CASE I. Both $|s|, |t| < 1$. Then (N,p) is Mercerian by Theorem 5.

CASE II. $|s| < 1 < |t|$. Let $A = (N,1-tz)$, $B = (N,1-sz)$. Then by Theorem 5 and Lemma 7, $c_{AB} = c \oplus y$ where $y = B^{-1}(\{t^n\})$. The easiest way to compute the last expression is to replace B by $(1-s)B$ and then replace b_{11} by 1. This results in a matrix which is equivalent for our purposes and has $(Bx)_n = x_n - sx_{n-1}$ and $(B^{-1}x)_n = \sum_{k=0}^{n} s^{n-k}x_k$. So $y_n = B^{-1}(\{t^n\})_n = \dfrac{s^{n+1}-t^{n+1}}{s-t}$; $y \in c \oplus \{t^n\}$ and $c_{AB} = c \oplus t^n$.

CASE III. Both $|s|,|t| > 1$, $s \neq t$. As in Case II, $c_{AB} = c_B \oplus y$ with $y = B^{-1}(\{t^n\}) \in c_B \oplus \{t^n\}$ so $c_{AB} = c_B \oplus \{t^n\} = c \oplus \{s^n\} + \{t^n\}$.

CASE IV. Both $|s| = |t| > 1$, $s = t$. As in Case II, $(B^{-1}x)_n = \sum_{k=0}^{n} s^{n-k}x_k$ so $y_n = B^{-1}(\{s^n\})_n = \sum_{k=0}^{n} s^{n-k}s^k = ns^n$; $c_{AB} = c_B \oplus y = c \oplus \{s^n\} \oplus \{ns^n\}$. This is exactly like the case of repeated roots in differential equations where a derivative ns^{n-1} occurs.

This then is the *Kubota-Petersen* theorem of 1952 which gives c_A explicitly if $A = (N,p)$ with $p(z)$ a polynomial with no zeros on $|z| = 1$. Namely $c_A = c \oplus a \oplus b \oplus\ldots$ where each of $a,b,c\ldots$ has the form $\{n^m\alpha^{-n}\}$ according as p has a zero α with $|\alpha| < 1$ of multiplicity m. Kubota's Theorem 5 was given in 1917.

8. THEOREM. *A polynomial matrix* (N,p) *is Tauberian if and only if* $p(z)$ *has no zeros in* $|z| = 1$.

This is by Lemma 4 and Kubota-Petersen.

9. EXAMPLE. The matrix $Q = (N,p)$ with $p(z) = 1 + z$ has a large convergence domain. None of the above theory applies to it.

Some more information is contained in [55], pp. 161-167. Also the article [48] is highly recommended.

CHAPTER 3
TRIANGLES AND BANACH SPACE

3.0. FUNCTIONAL ANALYSIS

1. *Hahn-Banach.* This takes many forms. The most useful is: if S is a vector subspace and $f = 0$ on S implies $f(x) = 0$, then $x \in \overline{S}$. If E is a set and $f = 0$ on E implies $f = 0$, then E is *fundamental* i.e. the span of E is dense.

Also: if X is a subspace of Y and f is a linear functional on X which is continuous in the relative topology of Y, then f can be extended to a continuous linear functional on Y. [80], Corollary 7-2-12, Theorem 7-2-11.

2. Two comparable complete metrizable topologies on a vector space must be equal. ([80], Corollary 5-2-7.)

3. The sum of a closed and a finite dimensional subspace must be closed. ([80], Theorem 6-3-3.)

4. Let p, q be non-equivalent norms with $q \leq Mp$. Then for each $\varepsilon > 0$ there exists x with $p(x) = 1$, $q(x) < \varepsilon$. Non-equivalence is guaranteed, for example, if p is complete and q is not. ([80], ##2-1-9, 2-2-1.)

5. An absolutely convergent series (in a complete space) is convergent. ([80], Theorem 5.2.2.)

3.1. HISTORICAL

Our subject began here. In 1927 S. Mazur won the prize of the University of Lwow for solving the consistency problem, Theorem 3.3.6. His matrices were regular triangles and his functional analysis was the brand new Banach space theory. To escape from triangles, Mazur and W. Orlicz and, independently, K. Zeller used Fréchet spaces; still later even the matrices were abandoned! We shall show this process, doing just a little of the special theory in order to motivate the more general.

NOTE. The results in §3.5 are for general conservative matrices, not just triangles.

3.2. CONVERGENCE DOMAIN

1. The general continuous linear function on c has the form $g(y) = \chi \lim y + ty$ where $t_k = g(\delta^k)$, $\chi = \chi(g) = g(1) - \Sigma t_k$, $t \in \ell$. (1.0.2).

Now if A is a triangle, $A : c_A \to c$ is one to one, linear, onto so c_A becomes a Banach space equivalent to c with the identification norm $\|x\|_A = \|Ax\|_\infty$.

2. THEOREM. *The coordinates are continuous on c_A (A a triangle); i.e. for each n, the map P_n is continuous. Here $P_n(x) = x_n$.*

Let $B = A^{-1}$, also a triangle. Then $|x_n| = |\Sigma\, b_{nk}(Ax)_k|$ $\leq (\sum_{k=1}^{n} |b_{nk}|).\|x\|_A$. Now see 1.0.1.

3. THEOREM (INCLUSION AND INVARIANCE). *Let A, B be triangles with $c_B \supset c_A$. Then the topology of c_A is larger than the relative topology of c_B. If A, B are equipotent the topologies are the same.*

Only the first statement needs to be proved. Let $C = BA^{-1}$. Then for $x \in c_A$, $\|x\|_B = \|Bx\|_\infty = \|CAx\|_\infty \leq \|C\|.\|x\|_A$. Also $\|C\| < \infty$ by 1.7.5 and 1.3.6.

4. THEOREM. *Let* A *be a triangle. Then the general* $f \in c_A'$ *has the form* $f(x) = \mu \lim_A x + t(Ax)$, $t \in \ell$.

Here t, x are thought of as a row and column vector, respectively, and $t(Ax) = \Sigma t_n(Ax)_n$. Let $g(y) = f(x)$ for $y = Ax$ so that $g \in c'$. Then by Remark 1, $f(x) = \chi \lim y + ty$ which gives the result. Note that $\mu(f) = \chi(g)$. It is not worthwhile at this point (although quite easy) to write a formula for $\mu(f)$ using Remark 1. (15.6.1).

5. THEOREM. *Let* A *be a conservative triangle and* $f \in c_A'$. *Then* $\chi(f) = \mu(f)\chi(A)$ *(1.3.8.)*

For $x \in c$, $f(x) = \mu \lim_A x + \gamma x$ where $\gamma = tA$, by Theorem 4 and 1.4.4. Thus $\chi(f) = \mu \cdot \chi(\lim_A) + \chi(u)$ where $u(x) = \gamma x$. It is trivial that $\chi(u) = 0$ and that $\chi(\lim_A) = \chi(A)$.

6. THEOREM. (INCLUSION AND INVARIANCE). *Let* A, B *be conservative triangles with* $c_B \supset c_A$. *If* A *is conull, so is* B. *If* A, B *are equipotent they are both coregular or both conull.*

Only the first statement needs to be proved. By Theorem 3, we may apply Theorem 5 to $\lim_B|c_A$ yielding $\chi(B) = \mu\chi(A)$ from which the result is obvious.

7. COROLLARY. *A conull triangle must sum divergent sequences. A Mercerian triangle must be coregular.*

The two statements are the same. The second is Theorem 6 with $B = I$.

3.3. THE PERFECT PART

For a conservative triangle A, c is a subspace of c_A. Its closure $\bar{c}$ in c_A is called the *perfect part* of c_A. If c is dense, A is called *perfect*. The Hahn-Banach theorem suggests considering those f which vanish on c.

1. THEOREM. *Let* A *be a coregular triangle,* $f \in c_A'$, $f = 0$ *on* c. *Then* $f(x) = t(Ax)$ *with* $tA = 0$, $t \in \ell$. *(The last two conditions are written* $t \perp A$*).*

First, $\chi(f) = 0$, hence by 3.2.5, $\mu(f) = 0$ and so $f(x) = t(Ax)$. Also $0 = f(\delta^k) = (tA)_k$ $(= \Sigma\, t_r a_{rk})$.

2. COROLLARY. *Let* A *be a coregular triangle. The perfect part of* c_A *includes all the bounded sequences in* c_A.

Let $f \in c_A'$, $f = 0$ on c. By Theorem 1 and 1.4.4. $f(x) = 0$ whenever $x \in c_A \cap \ell^\infty$. By the Hahn-Banach theorem (3.0.1) such $x \in \overline{c}$.

The conclusion of Corollary 2 may, but need not, hold for conull triangles. (13.5.1, 13.5.2).

The next definition is motivated by Theorem 1.

3. DEFINITION. *A matrix* A *is said to be of Type M if* $t \perp A$ *(Theorem 1) implies* $t = 0$.

4. THEOREM. *A coregular triangle is perfect if and only if it is of type M.*

If A is of type M it is perfect by Theorem 1 and the Hahn-Banach theorem. Conversely if A is perfect and $t \perp A$, let $f(x) = t(Ax)$ for $x \in c_A$. Then $f \in c_A'$ by 3.2.4, also $f = 0$ on c by 1.4.4. Since c is dense, $f = 0$ so $t = 0$ since t is uniquely determined by f in 3.2.4.

5. Sufficiency is false for conull; indeed a multiplicative 0 triangle *cannot* be perfect since $\lim_A \in c_A'$, $\lim_A = 0$ on c, but $\lim_A x = 1$ if x is the solution of $Ax = 1$. Such a matrix may be of type M, for example, take $(Ax)_n = x_n - x_{n-1}$. A conull triangle may be perfect (4.3.11); such a matrix must be of type M by the proof of Theorem 4.

We now give Mazur's famous consistency theorems.

6. THEOREM. *Let* A *be a regular triangle. (i)* A *is consistent with every regular matrix* B *such that* $c_B \supset c_A$ *iff* A *is of type* M. *(ii) For every regular matrix* B *with* $c_B \supset c_A$, A *is consistent with* B *on* $c_A \cap \ell^\infty$.

We shall prove instead a more general result:

7. THEOREM. *Let* A *be a coregular triangle. (i)* A *is consistent with every stronger matrix* B *such that* $\lim_B = \lim_A$ *on* c *iff* A *is of type* M. *(ii) For every stronger matrix* B *such that* $\lim_B = \lim_A$ *on* c, A *is consistent with* B *on* $c_A \cap \ell^\infty$.

Part (ii) and sufficiency in Part (i) will be trivial from Corollary 2 and Theorem 4, respectively as soon as it is shown that $\lim_B | c_A \in c_A'$. This follows from the formula $\lim_B x = \lim_n \lim_m \sum_{k=1}^{m} b_{nk}x_k$ for $x \in c_A$ applying 3.2.2 and the Banach-Steinhaus theorem (twice). To prove necessity in Part (i), suppose that A is not of type M. Let $t \perp A$, $t \neq 0$. Let $M = M(t)$, 1.8.12, and $B = MA$. Then $c_B = c_A$ by 1.7.16. Also by 1.4.4, $(Bx)_n = [M(Ax)]_n = (Ax)_n + \sum_{k=1}^{n-1} t_k(Ax)_k$ so, letting $n \to \infty$, $\lim_B x - \lim_A x = t(Ax)$. For $x \in c$ this is 0 as in the proof of Theorem 4. But if x is chosen so that $t(Ax) \neq 0$ then $\lim_B x \neq \lim_B x$; for example if $t_k \neq 0$ we may solve $Ax = \delta^k$ for x.

This result fails for conull triangles. The one given in Remark 5 shows this with $B = 2A$.

8. EXAMPLE. It is easy to check directly that Q (1.2.5) and $(C,1)$ are of type M, hence perfect. They also satisfy the criterion given in the next result.

9. THEOREM. *Let $A \in \Phi$ and suppose A has a right inverse A' with bounded columns. Then A is of type M.*

If $t \perp A$, then $t = t(AA') = (tA)A' = 0$ by 1.4.4 in which x is taken to be an arbitrary column of A'.

A very elegant application of these ideas is to the improvement of certain growth theorems "free of charge":

10. THEOREM. *Let A be a perfect triangle and suppose that $x_n = O(\lambda_n)$ for all $x \in c_A$ where $\lambda_n \to \infty$. Then $x_n = o(\lambda_n)$ for all $x \in c_A$.*

Let $f_n(x) = x_n/\lambda_n$. Then $\{f_n\}$ is a pointwise bounded sequence in c_A' by 3.2.2. By 1.0.5, $\{x: f_n(x) \to 0\}$ is a closed linear subspace of c_A. Since it includes c it includes all of c_A.

11. EXAMPLE. *For all $x \in c_Q$ (1.2.5) we have $x_n = o(n)$.* By 1.7.2 and Theorem 10, taking account of Example 8. *Exactly the same is true for $(C,1)$.* The extra little result of 1.7.2 becomes: $\lim(x_n - x_{n+2})$ exists implies $(x_n - x_{n+1})/n \to 0$.

12. EXAMPLE (J. DeFranza and D.J. Fleming) Let u be a real sequence with no zero terms and $u_n/u_{n-1} \to 1$. Let A be the matrix, clearly a coregular triangle (indeed $\frac{1}{2}A$ is regular), defined by $(Ax)_n = x_{n-1} + (u_n/u_{n-1})x_n$, $n = 1,2,\ldots$. Take $x_0 = 0, u_0 = 1$. (The matrix Q, 1.2.5, is the special case of $\frac{1}{2}A$ with $u_n = 1$.) Clearly $tA = 0$ if and only if t is a multiple of the sequence $\{1, -u_1, u_2, -u_3, \ldots\}$, hence *$A$ is of type M iff $u \notin \ell$.*

13. EXAMPLE. Let $A = (N,p)$ be a regular Nörlund matrix and suppose that $p(z) = 0$ for some z, $0 < |z| < 1$. Let $t_n = P_n z^n$. Then $t \in \ell$ by 2.5.10; also $(tA)_k = \sum_{n=k}^{\infty} t_n a_{nk} = \sum_{n=0}^{\infty} z^{n-k} p_n = z^{-k} p(z)$

= 0. It follows that A is not of type M. In particular A is not Mercerian (Theorem 4). This extends half of 2.6.5.

Source: [86].

3.4. WHEN c IS CLOSED

In contrast with 3.3.4, I do not know any internal test of a matrix A to see if c is closed in c_A; nevertheless the condition is important.

1. THEOREM. *Let $A \in \Delta$. If c is not closed in c_A, the perfect part (i.e. $\bar{c}$) must contain unbounded sequences. In particular A sums unbounded sequences.*

Let $X = \bar{c}$ and assume $X \subset \ell^\infty$. Then using 3.2.2, the sequence $\{P_n\}$ of coordinates is pointwise bounded and so $c = \{x \in X:\lim x_n$ exists$\}$ is closed in X. (1.0.5). Thus $c = \bar{c}$.

2. THEOREM. *Let $A \in \Delta$ and assume that c is closed in c_A. Then A is coregular and $\|\cdot\|_A$, $\|\cdot\|_\infty$ are equivalent on c.*

The two norms mentioned are complete and comparable, indeed $\|x\|_A \leq \|x\|_\infty \cdot \|A\|$. Hence they are equivalent (3.0.2). In c, $1 \notin \bar{c}_o$ where the closure is in $\|\cdot\|_\infty$; hence in c_A the same is true with the closure taken in $\|\cdot\|_A$ since the norms are equivalent on c. By the Hahn-Banach theorem there exists $f \in c_A'$ with $f = 0$ on c_o, $f(1) = 1$. Then $\chi(f) = 1$ so A is coregular. (3.2.5).

3. COROLLARY. *If a conservative triangle A sums a bounded divergent sequence, it must sum unbounded sequences.*

Suppose that $c_A \subset \ell^\infty$. Then c is closed in c_A by Theorem 1, and dense in c_A by Theorem 2 and 3.3.2. Hence $c_A = c$.

In contrast A may sum unbounded and no bounded divergent

sequences, (1.8.5).

4. THEOREM. *Let* $A \in \Delta$. *Then* c *is closed in* c_A *iff* A *is Tauberian.*

If c is closed, A is coregular by Theorem 2, and Tauberian by 3.3.2. Conversely, suppose that c is not closed. To construct a bounded divergent sequence in c_A we use the classical method of the *gliding hump:* If x^n has a hump in $[u_n, v_n]$ (i.e. $|x_k^n|$ is very small except when $u_n \leq k \leq v_n$, while x_k^n is 1 somewhere in this interval), and if $u_{n+1} > v_n$, it seems clear that $\Sigma\, x^n$ (in some sense) will be bounded and divergent. The details follow:

Fix an integer $u > 1$ and set $Z = \{x \in c_0 : x_1 = x_2 = \ldots = x_{u-1} = 0\}$. Then Z has codimension u in c, hence is not closed in c_A by 3.0.3. Thus $(Z, \|\cdot\|_A)$ is not complete. But $(Z, \|\cdot\|_\infty)$ is complete, so, by 3.0.4, for any $\varepsilon > 0$ there exists $x \in Z$ with $\|x\|_\infty = 1$, $\|x\|_A < \varepsilon$. (Note that $\|x\|_A \leq \|A\| \cdot \|x\|_\infty$). To summarize:

Given $\varepsilon > 0$ and integer $u > 1$, there exists x and integer $v > u$ such that

$$x_k = 0 \text{ for } 1 \leq k < u,\ |x_k| < \varepsilon \text{ for } k > v,\ \|x\|_\infty = 1 \tag{1}$$

and

$$x \in c_A,\ \|x\|_A < \varepsilon. \tag{2}$$

The existence of v is guaranteed since $x \in c_0$.

Choose x^1, v_1 to satisfy (1), (2) with $\varepsilon = 1/8$, $u_1 = 2$, $v_1 > u_1$; then x^2, v_2 with $\varepsilon = 1/16$, $u_2 = v_1 + 2$, $v_2 > u_2; \ldots$ $\ldots; x^n, v_n$ with $\varepsilon = 2^{-n-2}$, $u_n = v_{n-1} + 2$, $v_n > u_n$.

Now let $x = \Sigma\, x^n \in c_A$. The series converges since $\|x^n\|_A < \Sigma\, 2^{-n-2}$ (3.0.5), also $x_k = \Sigma_n\, x_k^n$ since coordinates are continuous (3.2.2).

We first prove that x *is bounded:* fix k; now $|x_k^n| < 2^{-n-2}$ if $v_n < k$, $x_k^n = 0$ if $u_n > k$ and this covers all values of n or all but one at which $|x_k^n| \le \|x^n\|_\infty \le 1$. Hence $|x_k| \le 1 + \Sigma\, 2^{-n-2}$.

Next we prove that x *is divergent:* choose k in one of intervals $[u_N, v_N]$ so that $x_k^N = 1$. Then $|x_k| \ge 1 - \Sigma\{|x_k^n| : k \ne N\}$ $\ge 1 - \Sigma\, 2^{-n-2} = 3/4$ and this is true for infinitely many k. But if $k = v_N + 1$, k is not in any $[u_n, v_n]$ and so $|x_k| \le \Sigma\, |x_k^n| \le \Sigma\, 2^{-n-2} = 1/4$ and this also is true for infinitely many k.

5. COROLLARY. *Let A be a conull triangle. Then A sums bounded divergent and unbounded sequences.*

By Theorem 2, c is not closed; by Theorem 4, A is not Tauberian. The rest is by Corollary 3.

3.5. BOUNDED SEQUENCES AND NON-TRIANGLES

The extension of the previous material to general conservative matrices and beyond matrices altogether requires using more general structures than Banach spaces. However the part that pertains to bounded sequences can be extended by a simple trick.

1. THEOREM. *Let $A \in \Gamma$ then there exists a multiplicative triangle M with $c_M \cap \ell^\infty = c_A \cap \ell^\infty$, $\chi(M) = \chi(A)$. Thus if A is coregular, M can be made regular; if A is conull, M is multiplicative - 0.*

First let $b_{nk} = a_{nk} - a_k$. Then $\|B\| \le \|A\| + \Sigma\, |a_k| < \infty$ by 1.3.7; $b_k = a_k - a_k = 0$ so B is multiplicative, $\chi(B) = \lim_B 1$ $= \chi(A)$. Also, for $x \in \ell^\infty$, $(Bx)_n - (Ax)_n = \Sigma\, a_k x_k$ for all n, so $c_B \cap \ell^\infty = c_A \cap \ell^\infty$.

Next, for each n choose m(n) so that $\sum_{k=m(n)}^{\infty} |b_{nk}| < 1/n$. Let $d_{nk} = b_{nk}$ for $k < m(n)$, 0 for $k \ge m(n)$. Then for $x \in \ell^\infty$,

$|(Bx)_n-(Dx)_n| < \|x\|_\infty/n \to 0$ so $c_D \cap \ell^\infty = c_B \cap \ell^\infty$ and $\lim_D = \lim_B$ on this set; D is also row-finite.

Next form a triangular matrix E by writing the first rows of the identity matrix followed by the first row of D when enough rows of the identity matrix have been written to keep the resulting matrix triangular. If the next row of D is no longer, write it immediately; otherwise repeat the first row of D as many times as ncecessary until the next row of D can be written to keep the resulting matrix triangular. Keeping this up we obtain a triangular matrix E with $c_E = c_D$ and $\lim_E = \lim_D$ on this set; E is triangular.

Finally if $e_{nn} = 0$ for any n replace it by $1/n$. The result is the required triangle M.

2. COROLLARY. *Let A be a conull matrix. Then A sums bounded divergent sequences. A Tauberian matrix must be coregular.*

This is by Theorem 1 and 3.4.5.

3. COROLLARY. *Let A, B be conull matrices. Then $c_A \cap c_B$ contains bounded divergent sequences.*

By Theorem 1 we may assume that A, B are multiplicative-0. Form a matrix D by writing the rows of A, B alternately. Then D is multiplicative-0 and so sums a bounded divergent sequence x by Corollary 2. Then $(Ax)_n = (Dx)_{2n}$ and $(Bx)_n = (Dx)_{2n+1}$ both tend to limits.

We are now able to improve 3.2.6 in several directions:

4. THEOREM (INVARIANCE). *Let A, B be conservative matrices with $c_B \supset c_A \cap \ell^\infty$. If A is conull so is B. If B is coregular so is A. If $c_B \cap \ell^\infty = c_B \cap \ell^\infty$, they are both coregular or both conull. The same holds a fortiori if $c_B \supset c_A$, $c_B = c_A$ respectively.*

Only the first statement needs to be proved. Let $D = B - \chi(B)I$. Then D is conull and by Corollary 3 there exists bounded divergent $x \in c_D \cap c_A$. By hypothesis, $x \in c_B$. Thus $\chi(B)x = Bx - Dx$ is convergent, so $\chi(B) = 0$.

5. COROLLARY. *No coregular matrix may sum all the bounded sequences.*

Apply Theorem 4 taking A to be any conull matrix, e.g. 0. This is the same as 1.7.21.

CHAPTER 4
FK SPACES

4.0. FUNCTIONAL ANALYSIS

1. A locally convex metrizable space has its topology defined by a sequence p of seminorms in the sense that $x \to 0$ iff $p(x) \to 0$ for each p. [80], #7-2-6, Theorem 7-2-2, Example 4-1-8.

2. The space ω of all sequences is a locally convex Fréchet space with the property that $x \to 0$ iff $x_n \to 0$ for each n; $f \in \omega'$ iff there exists $a \in \phi$ such that $f(x) = ax$. The topology is also defined by the sequence p of seminorms where $p_n(x) = |x_n|$ (See the preceding paragraph.) [80], Examples 5-5-2, 4-1-8, Theorem 7-2-2. A linear map $f : X \to \omega$ is continuous iff $P_n \circ f : X \to K$ is continuous for each n. [80], Remark 5-5-7.

3. If (X,p) is a space with topology determined by seminorms p, as in 1, $f \in X'$ iff there exist M and $p_1, p_2, \ldots, p_m$, selected from the seminorms p, such that $|f(x)| \leq M\Sigma\ p_k(x)$ for all x. [80], Theorem 7-2-5.

4. If $f : X \to Y$, the graph of f is $\{(x,f(x)) : x \in X\}$; it is a subset of $X \times Y$ and if it is closed, f is said to have *closed graph*. If X, Y are Fréchet spaces and f is linear and has closed graph, then f is continuous. This is the *closed graph theorem*. [80], Theorem 5-3-1. Any continuous map to a Hausdorff space has closed graph [80], Corollary 5-3-3.

5. ϕ, the space of finite sequences, has no Fréchet topology. [80], #6-4-107, Example 13-3-18.

6. If X, Y are Banach spaces, $B(X,Y)$, the continuous linear maps: $X \to Y$, is a Banach space. [80], Example 3-1-7.

7. A finite dimensional vector space X has a unique Hausdorff vector topology T; (X,T) is a Banach space. [80], Theorem 6-3-2.

8. Let p, q be seminorms on a vector space X and suppose that a linear functional f satisfies $|f(x)| \leq p(x) + q(x)$ for all x. Then there exist g, h with $f = g + h$, $|g(x)| \leq p(x)$, $h(x) \leq q(x)$. [80], Lemma 7-2-15.

9. Let a vector space X have a collection C of locally convex topologies. For each $T \in C$ let $P(T)$ be a set of seminorms which generates T as in 1. Let $P = \cup\{P(T) : T \in C\}$. Then P generates a topology on X called $\sup C$ with the property that $x \to 0$ iff $p(x) \to 0$ for all $p \in P(T)$ and all $T \in C$. Thus $x \to 0$ iff $x \to 0$ in (X,T) for each $T \in C$. If each $T \in C$ is a norm topology and C has finitely many members, $\sup C$ is given by a norm: the sum of these norms. [80] #7-2-2.

10. $f \in (X, \sup C)'$ (see 9) iff there exist $T_1, T_2 \ldots T_n \in C$ and $g_i \in (X,T_i)'$, $i = 1,2,\ldots,n$ such that $f = \Sigma\, g_i$. [80], Theorem 7-2-16.

11. A continuous linear map between topological vector spaces remains continuous when each space is given its weak topology. See 1.0.7 and [80], Example 11-1-4.

12. Every scalar homomorphism on a Banach algebra is continuous [79], Theorem 14.2.1; p. 277, line 9.

4.1. INTRODUCTION.

All the spaces met in summability (at least in this book) are sequence spaces i.e. linear subspaces of ω, the set of all sequences (of real or complex numbers.) So far they have been Banach spaces. This cannot continue. For $A = 0$, $c_A = \omega$ which, with its familiar topology, is not a normed space. (See 4.2.12) To clarify this last sentence, the various sequence spaces, such as c_A, will be given topologies which make them locally convex Fréchet spaces with the additional property that *coordinates are continuous*. We have seen this in 3.2.2. with applications in 3.3.6, 3.3.9, 3.4.1. A strong impetus for the development by the Polish mathematicians of this generalization of Banach space came directly from summability.

There are some exceptions to the above remarks — on occasion one deals with spaces of matrices or of operators as in §1.5; also there are fruitful generalizations to function spaces which we do not cover.

What is going to happen now is that spaces of the type just mentioned, called FK (Fréchet-Koordinat) spaces, will be studied briefly. It will then be shown that the summability spaces can be given this structure. This leads to a host of fascinating, even audacious, questions. Let P be some property which a matrix may have. Is P invariant? This usually means: if $c_A = c_B$ must A, B both or neither have property P?

Now suppose P is invariant. Then P is really a property of the space c_A rather than of the matrix A. This leads to a problem in the *naming program:* can P be defined in terms that apply to FK spaces and with no mention of a matrix? This has two effects: it automatically proves that P is invariant, and it allows us to discuss FK spaces with property P whether or not they arose from matrices.

We have seen one instance: type M; restricting ourselves entirely to coregular triangles the naming problem was solved in

3.3.4. Hence it is known (using 3.2.3, invariance of topology) that type M is invariant. The name is *perfect*. Now one can speak of perfect FK spaces.

The next beautiful possibility (the *FK Program*) is that some classical theorem for matrices has the form "P implies Q". Having solved the naming problem for P and Q (hence knowing that they are invariant) we make the audacious conjecture that every FK space with property P also has property Q. As an example, consider the fact that a type M coregular triangle must sum bounded divergent sequences by 3.3.4, 3.4.4. (We are excluding the trivial case $c_A = c$.) This is superseded by the Meyer-König and Zeller result (6.1.1), that any FK space in which c is not closed must contain bounded divergent sequences. My favorite example (partly because I had the honor of suggesting it to A. K. Snyder who did not realize that I was joking) is the supplanting of Agnew's proof that every multiplicative-0 matrix sums a large class (called Ω) of sequences by Snyder's proof (6.4.3) that every conull FK space includes Ω. (He, and independently, Jurimae, had first solved the naming problem for conull, 4.6.1).

4.2. FK SPACES

These spaces can be introduced with just a trifle more generality than is needed. This actually enhances understanding and in a very short time we shall specialize to sequence spaces. The more general setting applies to function spaces which arise in summability but will not be treated here; it also has applications in Banach algebra and elsewhere.

Let H be a topological vector (Hausdorff) space [80],

Definition 4.1.1. (A large part of the theory about to be presented uses much weaker assumptions on H, namely that it is a Hausdorff space and a vector space. This assumption covers this and the succeeding section for example)

An *FH space* is a locally convex Fréchet space X such that X is a vector subspace of H and the topology X is larger than the restriction to X of the H topology i.e. the inclusion map: $X \to H$ is continuous.

The reader may take $H = \omega$ throughout. The only casualty will be 14.1.8 in which H is the dual of a Banach space.

1. The theory of FH spaces can be developed without the assumption of local convexity. This is done in this section. However, since the spaces actually arising in this book are all locally convex we have chosen to incorporate this assumption into the definition.

If the phrase "let X, Y be FH spaces" is encountered it is assumed that H is the same for X, Y.

An *FK space* is an FH space with $H = \omega$, i.e. an FK space is a locally convex Fréchet space which is made up of sequences and has the property that coordinates are continuous as in 3.2.2. This part of the FK program is carried out by fiat.

A BH or BK space is the special case of the foregoing in which the Fréchet space is a Banach space.

The most prominent examples of BK spaces are c_0, c, ℓ, ℓ^∞; in each of these spaces $|x_n| \le \|x\|$, so that coordinates are continuous. The space ω is an FK space (4.0.2). If A is a triangle, c_A is a BK space by 3.2.2.

2. THEOREM. *Let X be a Fréchet space, Y an FH space and $f : X \to Y$ linear. If $f : X \to H$ is continuous, then $f : X \to Y$ is continuous.*

It is sufficient, by 4.0.4 to show that f has closed graph. Let T_H be the topology of H restricted to Y. Then $f:X\to(Y,T_H)$ has closed graph by 4.0.4. Now the graph of f is closed in $(X,T_X) \times (Y,T_H)$, hence in $(X,T_X) \times (Y,T_Y)$ since the latter topology is larger $(T_Y \supset T_H)$ i.e. has more closed sets.

3. COROLLARY. *Let X be a Fréchet space, Y an FK space and $f : X \to Y$ linear. If $P_n \circ f : X \to K$ is continuous for each n, then $f : X \to Y$ is continuous.*

The map $P_n : Y \to K$ is defined by $P_n(y) = y_n$ as usual. The given condition is equivalent to the continuity of $f : X \to \omega$ by 4.0.2 so the result follows by Theorem 2.

4. COROLLARY. *Let X, Y be FH spaces with $X \subset Y$. Then the topology of X is larger (on X) than the topology of Y. They are equal if and only if X is a closed subset of Y. In particular the topology of an FH space is unique i.e. there is at most one way to make a vector subspace of H into a FH space.*

Simply apply Theorem 2 to the inclusion map: $X \to Y$. This gives all the result except the statement about equality of the topologies. If X is closed in Y it becomes an FH space with $T_Y|X$; by uniqueness this is equal to T_X. Conversely if $T_Y|X = T_X$, X is a complete, hence closed, subset of Y.

This result explains a natural phenomenon: let X be any member of such a list as $\ell^{\frac{1}{2}}$, ℓ^2, ℓ^3, c_0, c, ℓ^∞, ω. The topology of X is seen to be larger the smaller X is; for example $\|x\|_2 = (\Sigma |x_i|^2)^{\frac{1}{2}} \geq \|x\|_\infty$ corresponding to $\ell^2 \subset c$. (Our inclusion of the non-locally convex space $\ell^{\frac{1}{2}}$ in the list is covered by Remark 1.)

Corollary 4 is part of the FK program in that it supersedes 3.2.3.

5. THEOREM. *Let* X, Y, Z *be* FH *spaces with* $X \subset Y \subset Z$ *and suppose that* X *is closed in* Z. *Then* X *is closed in* Y.

By hypothesis, X is closed in $(Y, T_Z|Y)$, hence in (Y, T_Y) by Corollary 4.

6. EXAMPLE. *If* X *is an* FK *space and* $c \subset X \subset \ell^\infty$, *then* c *is closed in* X since c is closed in ℓ^∞. The FK program is invoked in the equivalent statement: if c is not closed in X then X must contain unbounded sequences; see 3.4.1.

7. THEOREM. *Let* X, Y *be* FH *spaces with* $X \subset Y$. *Let* E *be a subset of* X. *Then* $cl_Y E = cl_Y cl_X E$. *In particular* $cl_X E \subset cl_Y E$.

$\supset$: $cl_Y E \supset cl_X E$ since $T_X \supset T_Y|X$ (Corollary 4). Take the closure of each side in Y. The opposite inclusion is trivial.

8. THEOREM. *Any matrix map between* FK *spaces is continuous.*

Let $A : X \to Y$ i.e $A \in (X:Y)$ (1.2.7). It is sufficient, by Corollary 3, to show that for each n, $x \to (Ax)_n$ is a continuous map: $X \to K$. Now $(Ax)_n = \lim_{m\to\infty} \sum_{k=1}^{m} a_{nk}x_k$ and each map $x \to \sum_{k=1}^{m} a_{nk}x_k$ is continuous since it is a finite linear combination of coordinates. The result follows by the Banach-Steinhaus theorem (1.0.4).

It seems that BK spaces are smaller than FK spaces which are not BK spaces. This is made precise in the next few results.

9. DEFINITION. *A sequence* t *is called a growth sequence for a set* S *of sequences if* $x_n = O(t_n)$ *for all* $x \in S$. *It is called a growth sequence for a matrix* A *if it is a growth sequence for* c_A.

It was shown in 1.7.1 that every triangle has a growth sequence. This is included in the following FK Program result.

10. THEOREM. *Every* BK *space* X *has a growth sequence.*

With $P_n(x) = x_n$ we have $P_n \in X'$ and so $\|P_n\| < \infty$. (1.0.1). Since for each $x \in X$, $|x_n| = |P_n(x)| \leq \|P_n\| \cdot \|x\|$ for all n, $\{\|P_n\|\}$ is a growth sequencc for X.

11. THEOREM. *Let* E *be a set of sequences. Then* E *is included in some* BK *space if and only if* E *has a growth sequence.*

Necessity is by Theorem 10. Now let t be a growth sequence for E. We may assume that $t_i \neq 0$ for all i since replacing any $t_i = 0$ by $t_i = 1$ preserves the property of being a growth sequence. Let $X = \{x : x_n = O(t_n)\}$ with $\|x\| = \|x/t\|_\infty$. Then X is isometric with ℓ^∞ under the map $x \to x/t$ so it is a Banach space. Also, for $x \in X$, $|x_n| \leq |t_n| \cdot \|x\|$ so X is a BK space. (1.0.1).

12. THEOREM. *No infinite dimensional closed subspace of* ω *is included in a* BK *space. In particular* ω *is not a* BK *space.*

Suppose that E is a closed subspace of ω, that X is a BK space, and $E \subset X$. Then E is closed in X by Theorem 5 so E is a BK space. By Corollary 4 the E norm is continuous in the ω topology on E and so for $x \in E$, $\|x\| \leq M \sum_{k=1}^{n} |x_k|$ using 4.0.3; M, n are independent of x. Hence if $x_1 = x_2 = \ldots = x_n = 0$ it follows that $x = 0$; thus the map $x \to (x_1, x_2, \ldots, x_n)$ is an isomorphism of E into R^n and so E is at most n-dimensional.

Of course the last part is immediate from Theorem 10.

The property given next plays a key role in summability.

13. DEFINITION. *An FK space X is said to have AK, or be an AK space, if $X \supset \phi$ and $\{\delta^n\}$ is a basis for X, i.e. for each x, $x^{(n)} \to x$, where $x^{(n)}$, the nth section of x is $\sum_{k=1}^{n} x_k\delta^k$; otherwise expressed, $x = \Sigma x_k\delta^k$ for all $x \in X$. The space is said to have AD, or be an AD space if ϕ is dense in X.*

The initials come from Abschnitts-Konvergenz, sectional convergence, and Abschnitts-dicht, section dense.

14. EXAMPLE. *The spaces l and cs have AK.* By cs we mean the space of convergent series, i.e $\{x:\Sigma x_k$ converges$\}$. It is c_A, where $(Ax)_n = \sum_{k=1}^{n} x_k$; A is a triangle so c_A is a BK space, (3.2.2), with $\|x\|_{cs} = \sup|\sum_{k=1}^{n} x_k|$. AK follows from the calculation $\|x-x^{(n)}\|_A = \sup_m|\sum_{k=n+1}^{m} x_k| \to 0$. An even easier calculation shows that ℓ has AK.

15. THEOREM. *The intersection of countably many FH spaces is an FH space. If each X_n is an FK space with AK then $X = \cap X_n$ has AK.*

Let T_n be the topology on X_n. Place on X the topology $T = \sup\{T_n|X_n\}$. (4.0.9). The coordinates are continuous in each T_n, hence in the larger topology T. If x is a T Cauchy sequence in X it is a T_n Cauchy sequence in X_n for each n, hence convergent; say $x \to t_n \in X_n$. Then for each n, $x \to t_n$ in H so all the t_n are the same; say $t_n = t$. Clearly $t \in X$ and $x \to t$ in T_n for each n, so $x \to t$ in X by 4.0.9. For the second statement, observe that $x^{(k)} \to x$ in each X_n hence in X by 4.0.9.

16. If there are only finitely many spaces in Theorem 15 the metric on X may be taken to be the sum of the metrics on the X_n. (4.0.9). In particular *the intersection of finitely many BH spaces is a BH space.*

17. Any sequence space which is a Banach algebra is automatically a BK space, since the coordinates, being scalar homomorphisms, are continuous. (4.0.12).

4.3. CONSTRUCTION

Every locally convex metrizable space X has its topology defined by a finite or infinite sequence of seminorms which we write p or $\{p_n\}$ (4.0.1). What this means is that if $\{x^k\}$ is a sequence in X, then $x^k \to 0$ iff $p_n(x^k) \to 0$ for each n. For example $(\omega, |P|)$ is an FK space where $P = \{P_n\}$ is the sequence of coordinates and $|P| = \{|P_n|\}$, a sequence of seminorms; $x^k \to 0$ in ω iff $x_n^k \to 0$ for each n (4.0.2). Another example is c in which $p(x) = \|x\|_\infty$; there is only one seminorm, a norm in this case, and $x^k \to 0$ in c iff $\|x^k\|_\infty \to 0$.

The notation (X,p), then, refers to a vector space X and a sequence p of seminorms — the metrizable topology resulting has the convergent sequences just mentioned.

The theory of FK spaces turns out to apply to all convergence domains. We begin with a basic construction:

1. THEOREM. *Let (X,p), (Y,q) be FK spaces and A a matrix defined on X i.e. $X \subset \omega_A$. Let $Z = X \cap Y_A = \{x \in X : Ax \in Y\}$. Then Z is an FK space with $p \cup (q \circ A)$.*

This means that Z is given all the seminorms $p_1, p_2, \ldots$ and $q_1 \circ A$, $q_2 \circ A \ldots$. Since this is a countable set which includes p it yields a metrizable topology larger than that of X, hence of ω. It remains to show that it is complete. Let x be a Cauchy sequence in Z. Then x is Cauchy in X, say $x \to t$ in X i.e. $p_n(x-t) \to 0$ for each n. Also Ax is Cauchy in Y since $q_n(Ax) = (q_n \circ A)(x)$; say $Ax \to b$ in Y. Then $Ax \to At$ in ω since $A : X \to \omega$ is continuous by 4.2.8, and $Ax \to b$ in ω since Y is an FK space. Thus $b = At$. Hence $t \in Z$ and $x \to t$ since $p_n(x-t) \to 0$ and $(q_n \circ A)(x-t) = q_n(Ax-b) \to 0$ for each n.

The preceding proof may just as well have been written with H instead of ω. To take advantage of the special character of ω we might have written the last few steps as: $(Ax)_k \to (At)_k$ for each k and $(Ax)_k \to b_k$ for each k hence $(At)_k = b_k$ for each k so that $b = At$. Hence $t \in Z$, etc.

2. THEOREM. *If in Theorem 1, A is one to one on X and $A[X] \supset Y$, then the seminorms p can be omitted i.e. Z is an FK space with $q \circ A$.*

It is trivial that Z is a Fréchet space with $q \circ A$, indeed it is linearly isometric with Y under the map $A : X \to Y$. Thus using Theorem 1, Z has two comparable Fréchet metrics; they must be equal by 3.0.2.

3. THEOREM. *Let A be a row-finite matrix. Then (c_A, p) is an FK space where $p_0 = \|\cdot\|_A$, $p_n = |x_n|$ for $n = 1,2,\ldots$. If A is a triangle, (c_A, p_0) is a BK space.*

We apply Theorem 1 with $X = \omega$, $Y = c$, so that $Z = c_A$. The seminorms of X are p_n, $n = 1,2,\ldots$; Y is a BK space with $q = \|\cdot\|_\infty$ so $q \circ A = \|\cdot\|_A$. The last part is by Theorem 2 or the more elementary considerations in §3.2

To extend this to arbitrary matrices we introduce the important notion of *attachment:*

4. DEFINITION. *Let* Y *be an FK space and* z *a sequence. Then* $z^{-1}\cdot Y = \{x \in \omega : x\cdot z \in Y\}$ *where* $x\cdot z = \{x_n z_n\}$.

Note that the definition makes sense even if z^{-1} does not; e.g. $0^{-1}\cdot Y = \omega$. Contrast $x\cdot z$ with xz which means $\Sigma\, x_n z_n$.

As a special case we obtain a *Köthe-Toeplitz dual:*

5. DEFINITION. *The space* $z^{-1}\cdot cs$ *is written* z^β; Z^β *is defined to be* $\cap\{z^\beta : z \in Z\} = \{x : \Sigma\, x_n z_n$ *is convergent for all* $z \in Z\}$.

6. THEOREM. *Let* (Y,q) *be an FK space and* z *a sequence. Then* $z^{-1}\cdot Y$ *is an FK space with* $p \cup h$ *where* $p_n = |x_n|$ *and* $h(x) = q(z\cdot x)$. *If* Y *has AK then* $z^{-1}\cdot Y$ *has AK also.*

Let $a_{nn} = z_n$, $a_{nk} = 0$ if $n \neq k$, and apply Theorem 1 with $X = \omega$. To prove the last part, $p_n(x-x^{(m)}) = 0$ for $m > n$, while $h_n(x-x^{(m)}) = q_n[z\cdot(x-x^{(m)})] = q_n[z\cdot x-(z\cdot x)^{(m)}] \to 0$.

7. THEOREM. *Let* z *be a sequence. Then* (z^β, p) *is an AK space with* $p_o(x) = \sup_m|\sum_{k=1}^m z_k x_k|$, $p_n(x) = |x_n|$. *For any* k *such that* $z_k \neq 0$, p_k *may be omitted. If* $z \in \phi$, p_o *may be omitted. (See Remark 10.)*

The first sentence follows from Theorem 6, Definition 5 and 4.2.14. The form of p_o can also be gotten from Theorem 3 with $(Ax)_n = \sum_{k=1}^n x_k z_k$. If $z_k \neq 0$, then, with the matrix A just mentioned, $|x_k| = |(Ax)_k-(Ax)_{k-1}|/|z_k| \leq 2p_o(x)/|z_k|$ and so p_k is redundant. If $z \in \phi$, $z^\beta = \omega$.

8. THEOREM. *Let* A *be a matrix. Then* $(\omega_A, p \cup h)$ *is an AK space with* $p_n(x) = |x_n|$, $h_n(x) = \sup_m|\sum_{k=1}^m a_{nk}x_k|$. *For any* k *such that the* kth *column of* A *has at least one non-zero term,*

p_k may be omitted. For any n such that the nth row of A has finite length, h_n may be omitted. (See Remark 10.)

Since $\omega_A = \cap\{r^\beta : r$ is a row of $A\}$, the first part is immediate from Theorem 7, and 4.2.15. If $a_{nk} \neq 0$, $|x_k| \leq 2h_n(x)/|a_{nk}|$ exactly as in the proof of Theorem 7 so p_k is redundant. If the nth row has finite length, h_n does not appear by the last part of Theorem 7.

9. Of course Theorem 8 could be expressed as a statement about Z^β (Definition 5) where Z is a countable set. Countability is essential, for example $\ell^\beta = \ell^\infty$ which does not have AK, and $\omega^\beta = \phi$ which is not an FK space at all (4.0.5). In Theorem 15 we examine certain non-countable intersections.

10. Let A be the identity matrix. Then apparently Theorem 8 says that *all* the seminorms are redundant for $\omega_A = \omega$. What Theorem 8 says is that each p_n is redundant if the h_m are present. Actually $h_m = p_m$ and so the first list of seminorms is just $p_1, p_2, \ldots$.

11. EXAMPLE. *A perfect conull triangle.* Let $a \in \ell$ with $a_k \neq 0$ for all k, and $(Ax)_n = \sum_{k=1}^{n} a_k x_k$. Then $c_A = a^\beta$ has AK by Theorem 7, hence ϕ is dense.

12. THEOREM. *Let (Y,q) be an FK space and A a matrix. Then $Y_A = \{x : Ax \in Y\}$ is an FK space with $p \cup h \cup q\circ A$ where p, h are as in Theorem 8. For any k such that the kth column of A has at least one non-zero term, p_k may be omitted. For any n such that the nth row of A has finite length, h_n may be omitted. (See Remark 10). If A is a triangle use only $q\circ A$.*

Apply Theorem 1 with $X = \omega_A$, an FK space by Theorem 8.

Then $Z = Y_A$ and the seminorms may be read off from Theorems 1 and 8. The remaining parts follow from Theorem 8 and the fact that if A is a triangle the map $A : Y_A \to Y$ is an equivalence.

13. THEOREM. *Let A be a matrix. Then c_A is an FK space with $p \cup h$ where $p_0 = \|\cdot\|_A$, p_n, h_n are as in Theorem 8 for $n = 1,2,\ldots$. For any k such that the kth column of A has at least one non-zero term, p_k may be omitted. For any n such that the nth row of A has finite length h_n may be omitted. (See Remark 10.)*

This is a special case of Theorem 12. See also Theorem 3.

14. THEOREM. *If X is a closed subspace of Y, then X_A is a closed subspace of Y_A.*

Define $f : Y_A \to Y$ by $f(y) = Ay$, a continuous map. Then $X_A = f^{-1}[X]$ is closed.

We now discuss a general form of dual space, called the *multiplier space*, which specializes to various Köthe-Toeplitz duals.

15. THEOREM. *Let X, Y be BK spaces with $X \supset \phi$. Let $Z = M(X,Y) = \cap\{x^{-1}\cdot Y : x \in X\} = \{z : x\cdot z \in Y$ for all $x \in X\}$. Then Z is a BK space.*

The space $B = B(X,Y)$ of all continuous linear maps from X to Y is a Banach space (4.0.6). Each member $z \in Z$ yields a diagonal matrix map $\hat{z} : X \to Y$, $\hat{z}(x) = x\cdot z$, which is continuous by 4.2.8. This clearly embeds Z in B. [If $\hat{z} = 0$, $z_n = \hat{z}(\delta^n) = 0$ so $z = 0$.] The induced norm is $\|z\| = \sup\{\|x\cdot z\| : \|x\| \leq 1\}$. To see that coordinates are continuous, fix n. Let $u = 1/\|\delta^n\|_X$, $v = \|\delta^n\|_Y$. Then $\|z\| \geq \|(u\delta^n)\cdot z\| = u\|z_n\delta^n\| = uv|z_n|$. It remains to show that Z is a closed subspace of B. Let $\hat{z} \to T \in B$. Then $\hat{z}(x) \to T(x)$ for each x and, since Y is an FK space,

$[\hat{z}(x)]_k \to T(x)_k$ for each k, i.e. $x_k z_k \to T(x)_k$. With $x = \delta^k$ this gives $z_k \to T(\delta^k)_k$ which we shall write as t_k. Thus $x_k z_k \to x_k t_k$ and, as just proved, $x_k z_k \to (Tx)_k$. Hence $Tx = x \cdot t$ so $T = \hat{t}$.

16. EXAMPLE. $M(X, cs) = X^\beta$ so Theorem 15 shows that *X^β is a BK space if X is.* Further

$$\|u\|_\beta = \sup\{\|u \cdot x\|_{cs} : \|x\| \le 1\}$$
$$= \sup\{|\sum_{k=1}^n u_k x_k| : n = 1, 2, \ldots; \|x\| \le 1\}.$$

17. EXAMPLE. Other Köthe-Toeplitz duals are also special cases of Theorem 15. They are the α-dual, $M(X, \ell) = X^\alpha$, and the γ-dual, $M(X, bs) = X^\gamma$, where *bs is the space of bounded series* i.e. $y \in bs$ iff

$$\|y\|_{bs} = \sup|\sum_{k=1}^n y_k| < \infty; \; bs = (\ell^\infty)_A \quad \text{where} \quad (Ax)_n = \sum_{k=1}^n x_k,$$

so it is a BK space with the norm just shown, by Theorem 12. Further

$$\|u\|_\gamma = \sup\{\|u \cdot x\|_{bs} : \|x\| \le 1\} = \sup\{|\sum_{k=1}^n u_k x_k| : n = 1, 2, \ldots; \|x\| \le 1\}$$

18. THEOREM. *Let X be a BK space. Then X^β is a closed subspace of X^γ.* For the norms are the same, as shown in the preceding examples.

As pointed out in Remark 9, Theorem 15 fails if X is not a BK space; Z need not even be an FK space.

Information about the spaces $M(X, Y)$ may be found in [55], pp. 11, 12, 122, 133.

4.4 DUAL SPACE

1. THEOREM. *Let* (X,p), (Y,q) *be FK spaces and* A *a matrix defined on* X *i.e.* $X \subset \omega_A$. *Let* $Z = X \cap Y_A = \{x \in X : Ax \in Y\}$. *Then* Z *is an FK space and* $f \in Z'$ *iff* $f = F + g \circ A$ *with* $F \in X'$, $g \in Y'$.

The first part is a repetition of 4.3.1. If f has the given form it is continuous since F and $g \circ A$ are continuous on Z with smaller topologies, respectively p and $q \circ A$. Conversely, let $f \in Z'$. By 4.0.3 and 4.0.8, $f = F + u$ with $|F(x)| \leq p'(x)$, $|u(x)| \leq q'(Ax)$ where F, u are linear functionals on Z, p', q' are positive linear combinations of finitely many p_n, q_n respectively. We may assume $F \in X'$ by the Hahn-Banach theorem (3.0.1). Define g on $A[X] \cap Y$ by setting $g(y) = u(x)$ whenever $y = Ax$. To see that g is well-defined, suppose that $y = Ax_1 = Ax_2$; then $|u(x_1) - u(x_2)| \leq q'[A(x_1 - x_2)] = 0$. Since $|g(y)| \leq q'(y)$ we may extend g to all of Y by the Hahn-Banach theorem.

2. THEOREM. *Let* Y *be an FK space and* A *a matrix. Then* Y_A *is an FK space and* $f \in Y'_A$ *iff* $f(x) = \alpha x + g(Ax)$ *for* $x \in Y_A$, *where* $\alpha \in \omega_A^\beta$ *and* $g \in Y'$ *(see 4.3.5 for the* β*-dual). If* A *is a triangle,* α *may be taken to be* 0.

In Theorem 1 take $X = \omega_A$, an AK space by 4.3.8. Then $F(x) = F(\Sigma x_k \delta^k) = \Sigma F(\delta^k) x_k = \alpha x$. Sufficiency is by the Banach-Steinhaus Theorem (1.0.4) since $\alpha \in \omega_A^\beta \subset Y_A^\beta$, and the fact that Y_A is an FK space. If A is a triangle the map $A : Y_A \to Y$ is an equivalence.

3. THEOREM. *Let* A *be a matrix. Then* $f \in c'_A$ *iff* $f(x) = \mu \lim_A x + t(Ax) + \alpha x$ *where* $t \in \ell$, $\alpha \in c_A^\beta$.

This is a special case of Theorem 2 using the representation of c'. (1.0.2.)

4. In this theorem we can replace c_A^β by ω_A^β. Sufficiency is true *a fortiori* and necessity is exactly what was proved in Theorem 2. There are strong reasons for giving the result as we did, bound up in questions of uniquess and invariance (see 5.4.12). To see the difference between the requirements on α, we know that if A is a triangle we may take $\alpha = 0$ in Theorem 3, (3.2.4), moreover if we insist that $\alpha \in \omega_A^\beta$ then we must have $\alpha \in \phi$ since $\omega_A^\beta = \omega^\beta = \phi$. However, consider the matrx A defined by $(Ax)_n = \sum_{k=1}^{n} a_k x_k$ where $a_k \neq 0$ for all k. Then $\lim_A x = \Sigma\, a_k x_k$ so we have a representation for this function with $\mu = 1$, $t = \alpha = 0$ and another with $\mu = t = 0$, $\alpha = a \notin \phi$. In the latter case $\alpha \in c_A^\beta = a^{\beta\beta}$ but $\alpha \notin \omega_A^\beta = \phi$. In applications of Theorem 3 we shall use particular representations for f as appropriate at the time.

5. THEOREM. *If A is row-finite we may take $\alpha \in \phi$ in Theorem 3; if A is a triangle we may take $\alpha = 0$. If $A \in \Phi$ we may take $\alpha \in \ell$.*

For as given in the proof of Theorem 2 we may choose $\alpha \in \omega_A^\beta = \omega^\beta = \phi$. The second part was covered in Remark 4. For the third part, as in Remark 4, take $\alpha \in \omega_A^\beta \subset c^\beta = \ell$.

6. In Remark 4 it was shown that μ is not always uniquely determined by f. A matrix is called *μ-unique* if all representations of each $f \in c_A'$ as in Theorem 3 have the same μ. Clearly if some one f has unique μ then they all do.

7. THEOREM. *Let A be a conservative matrix, $f \in c_A'$. Let f be represented as in Theorem 3. Then $\chi(f) = \mu \cdot \chi(A)$.*

First $a = \{a_k\} \in \ell$ where $a_k = \lim_n a_{nk}$ by 1.3.7, also $\alpha \in c_A^\beta \subset c^\beta = \ell$ (given in the proof of 1.3.2) so

$$\chi(f) = f(1) - \Sigma f(\delta^k)$$

$$= \mu \lim_A 1 + t(A1) + \Sigma\alpha_k - \Sigma_k(\mu a_k + \Sigma_n t_n a_{nk} + \alpha_k)$$

$$= \mu\, \chi(A) + t(A1) - (tA)1$$

and the result follows by 1.4.4 (ii).

8. COROLLARY. *A coregular matrix is* μ-*unique.*

At this point in §3.2 we deduced invariance of conull and coregular for triangles from the formula above (3.2.6). There is no need for this now since a much more general result was established in 3.5.4. (Deduction of the more special result from Theorem 7 is easier in that it does not use 3.4.4).

9. THEOREM. *Let* A *be a conservative matrix and* $f \in c_A'$. *Then there exists a sequence* γ *such that* $f(x) = \mu \lim_A x + \gamma x$ *for all bounded* $x \in c_A$.

This follows from Theorem 3 and 1.4.4 (ii) with $\gamma = tA + \alpha$.

10. THEOREM. *Let* Y *be an FK space and* z *a sequence. Let* $X = z^{-1}\cdot Y$. *Then* X *is an FK space and* $f \in X'$ *iff* $f(x) = \alpha x + g(z\cdot x)$, $\alpha \in \phi$, $g \in Y'$.

Let $a_{nn} = z_n$, $a_{nk} = 0$ if $n \neq k$ and apply Theorem 2; $\alpha \in \omega_A^\beta = \omega^\beta = \phi$.

4.5 COMPLEMENTS

In preceding sections FK spaces were built up by starting with ω and c and using the basic construction of 4.3.1 to build new FK spaces as in 4.3.3. Other construction methods were intersection (4.2.15), and the use of multipliers (4.3.15). Another important method of building new spaces is by adding spaces. It will be sufficient for our purposes to add just two spaces which have zero intersection.

In the next result continuity of addition in H is used for the first time.

1. THEOREM. *Let (X,p), (Y,q) be FH spaces with zero intersection i.e. $X \cap Y = \{0\}$. Let $Z = X + Y = \{x+y : x \in X, y \in Y\}$. Then Z is an FH space and X, Y are closed in Z. If X, Y are BH spaces, Z is a BH space.*

Each $z \in Z$ can be written uniquely in the form $x+y$ and we define $r(z) = p(x) + q(y)$. If z is a Cauchy sequence in (Z,r) then x, y are Cauchy sequences in (X,p), (Y,q), hence convergent; say $x \to s \in X$, $y \to t \in Y$. Let $u = s+t$. Then $r(z-u) = p(x-s) + q(y-t) \to 0$ i.e. $z \to u$ in (Z,r) proving that *this space is complete.* To see that *it is an FH space* suppose $z \to 0$. Let $z = x + y$ then $p(x) \leq p(x) + q(y) = r(z)$ so $x \to 0$ in X, hence in H; similarly $y \to 0$ in H and so $z = x + y \to 0$ in H. Next, *X is closed:* for $x \in X$, $r(x) = p(x) + q(0) = p(x)$ so X has the relative topology of Z, hence is closed by 4.2.4. The last statement is trivial.

2. COROLLARY. *Let Z be an FH space and X, Y algebraically complementary subspaces of Z i.e. $Z = X + Y$, $X \cap Y = \{0\}$. Suppose that X, Y can be given FH topologies. Then X, Y are closed in Z.*

By Theorem 1, Z can be given an FH topology in which its subsets X, Y are closed. By the uniqueness theorem (4.2.4) this must be the original topology on Z.

3. COROLLARY. *Let X, Z be FH spaces with $X \subset Z$. Suppose X has finite codimension in Z. Then X is closed in Z.*

The hypothesis means that Y in Corollary 2 has finite dimension. Let Y have the relative topology of H. By the uniqueness theorem (4.0.7), this makes Y a BH space and the result follows by Corollary 2.

4. EXAMPLE. *Any algebraic complement of* ℓ *in* c_o must be infinite dimensional, indeed it *cannot be made into an FK space* by Corollary 2, since ℓ is not closed in c_o. (The same is true for c_o, ℓ^∞ because c_o has no closed complement. We make no use of this. The result may be found in [80], Example 14-7-8.)

5. EXAMPLE (IMPORTANT). Let X be an FH space and $y \in H\backslash X$. Let Y be the (one dimensional) span of y. Then $X + Y$ is written $X \oplus y$. By Theorem 1 it is an FH space and X is a closed subspace.

6. THEOREM. *Let* Z *be a conservative FK space i.e.* $Z \supset c$. *Suppose that* c *is not closed in* Z. *Then* Z *contains infinitely many unbounded sequences which are linearly independent (mod* ℓ^∞*) i.e. every non-trivial linear combination is unbounded.*

Suppose the conclusion is false. Let $X = Z \cap \ell^\infty$, an FK space by 4.2.15. By Corollary 3 X is closed in Z. Now c is closed in X (4.2.6), hence closed in Z.

4.6 COREGULAR AND CONULL

We have seen that coregular and conull are invariant properties. (3.5.4.) We now solve the naming problem for these properties. (A.K. Snyder and E. Jurimae).

1. THEOREM. *Let* A *be a conservative matrix. Then* A *is conull if and only if* $1^{(n)} \to 1$ *weakly in* c_A.

The notation $1^{(n)}$ was given in 4.2.13. The condition is that $\Sigma f(\delta^k) = f(1)$ i.e. $\chi(f) = 0$, for all $f \in c_A'$. Necessity is by 4.4.7. Conversely, the condition implies in particular, taking $f = \lim_A$, that $\chi(A) = \chi(\lim_A) = 0$.

2. DEFINITION. *A conservative FK space X is called conull if $1^{(n)} \to 1$ weakly i.e. $\chi(f) = 0$ for all $f \in X'$, otherwise coregular.*

So a conservative matrix A is conull if and only if c_A is conull. It goes (almost) without saying that conull and coregular are invariant properties since the topology is invariant (4.2.4). This, and the next result, are parts of the FK program for 3.2.6.

3. THEOREM. *Let X, Y be FK spaces with $X \subset Y$. If X is conull so is Y. If X is coregular and closed in Y then Y is coregular.*

Since the inclusion map: $X \to Y$ is continuous (4.2.4), it remains continuous when X, Y have their weak topologies (4.0.11). This implies the first assertion. If X is closed in Y the map is a homeomorphism into, (4.2.4) hence its inverse is weakly continuous (4.0.11). If Y is conull this makes X conull.

Next comes the FK program for the first part of 3.4.2. (For the second part it is 4.2.4).

4. COROLLARY. *If c is closed in Y, Y is coregular.*

For $c = c_I$ is coregular. In particular ℓ^∞ is coregular.

We saw (3.2.7, 3.5.2) that a conull matrix must sum divergent sequences. The FK program for this is simply the statement that c is coregular. But we can collect a much better result, and, in Theorem 6 an even better one.

5. THEOREM. *A conull FK space must contain infinitely many unbounded sequences which are linearly independent (mod* ℓ^∞*).*

This is from 4.5.6 and Corollary 4.

6. THEOREM. *The intersection of countably many conull spaces is conull.*

Let $f \in X'$ where $X = \cap X_n$. (4.2.15) Then $f = \sum_{k=1}^{m} g_k$ where $g_k \in X_n'$ for some n. (4.0.10). It follows that $\chi(f) = \Sigma\, \chi(g_k) = 0$.

We saw (3.4.3) that if a conservative triangle sums bounded divergent sequences it also sums unbounded sequences. This is extended in Corollary 8 to an arbitrary conservative matrix. We begin with the extension of 3.3.2. The *perfect part* of any conservative FK space is $\bar{c}$, and a *perfect space* is one in which c is dense.

7. COROLLARY. *Let* A *be coregular. The perfect part of* c_A *includes all the bounded sequences in* c_A.

Let $f \in c_A'$, $f = 0$ on c. Then $\mu(f) = 0$ by 4.4.7 and $f(x) = \gamma x$ for bounded $x \in c_A$ by 4.4.9. Taking $x = \delta^k$ yields $\gamma_k = f(\delta^k) = 0$ so $f = 0$ on $c_A \cap \ell^\infty$. The Hahn-Banach theorem gives the result.

The conclusion of Corollary 7 may or may not hold for conull matrices (13.5.1, 13.5.2).

8. COROLLARY. *If* A *conservative matrix sums a bounded divergent sequence it must sum unbounded sequences.*

Suppose that $c_A \subset \ell^\infty$. Then c is closed in c_A by 4.2.6 and dense in c_A by Corollaries 4 and 7. (Other proofs: 6.5.6, 13.4.7).

9. EXAMPLE. *The FK program fails for Corollaries 7 and 8.* For example ℓ^∞ is coregular and its perfect part is c. Also an

FK space strictly between c and ℓ^{∞} can be obtained by taking $X = c \oplus z$, $z \in \ell^{\infty}$ by 4.5.5. The proofs of the results must use matrices in an essential way.

4.7 THE FK PROGRAM

With a few exceptions all earlier matrix results have been given their FK formulation or else it was shown that the formulation is invalid. From this point on all results will be given for FK spaces when possible — when a matrix result is given one may assume that the FK program fails or has unknown status.

The reader should be aware, however, that many of these results were discovered first for matrices. Since this is a summability book, our primary motivation is in matrices so we resist the temptation to develop FK theory per se.

Bennett and Kalton pointed out that the FK program can be applied more widely than was suspected — namely they observed that convergence domains are separable and so it is permissible to investigate separable FK spaces. This works well — details are given in Chapter 16, especially 16.2.6.

CHAPTER 5
REPLACEABILITY AND CONSISTENCY

5.0. FUNCTIONAL ANALYSIS

1. Let f be a linear functional on a topological vector space, $M = f^{\perp}$. Then M is a maximal subspace; it is either closed or dense; f is continuous iff M is closed. [80], Theorem 4-5-10.

2. Let E be a convex set in a locally convex space (X,t). The weak closure of E is the same as its t closure. [80], Corollary 8-3-6.

3. Every continuous linear map: $\omega \to \omega$ is range closed. [80], Theorem 12-4-20. This result is used only in 5.5.1 for which an elementary proof is also cited.

5.1. FUNCTION AS MATRIX

Throughout this chapter we use the fact that each $f \in c_A'$ has a representation

$$f(x) = \mu \lim_A x + t(Ax) + \alpha x, \quad t \in \ell, \ \alpha \in c_A^{\beta} \tag{1}$$

as given in 4.4.3. From (1) it follows that $c_A \subset \alpha^{\beta}$.

1. THEOREM. *If f has a representation (1) with $\mu \neq 0$ then there exists B with $c_B = c_A$, $\lim_B = f$.*

Note that f may also have a representation with $\mu = 0$. (4.4.4)

We may assume that $\mu = 1$. Let $D = MA$ where $M = M(t)$, Mazur's Mercerian matrix, 1.8.12. Let $b_{1k} = \alpha_k$, $b_{nk} = \alpha_k + d_{n-1,k}$

for $n > 1$. For $x \in c_A$ we have $M(Ax) = Dx \in c$ by 1.4.4 and since M is conservative. Thus $x \in c_B$ since $x \in c_A \subset \alpha^\beta$ and $(Bx)_n = \Sigma\, \alpha_k x_k + (Dx)_{n-1}$.

Also if $x \in c_B$ then $x \in \alpha^\beta$ since $(Bx)_1 = \Sigma\, \alpha_k x_k$. Hence $(Dx)_{n-1} = (Bx)_n - \Sigma\, \alpha_k x_k$ and so $Bx \in c$. We have now proved

$$c_A \subset c_B \subset c_D \tag{2}$$

Further $\lim_B x = \lim_D x + \alpha x = \lim_M Ax + \alpha x = f(x)$. The proof is concluded by noting that $c_D = c_A$ by 1.7.16.

2. *If* $\mu = 0$, *Theorem 1 holds with the weaker conclusion* $c_B \supset c_A$. Take $(Mx)_n = \sum_{k=1}^{n-1} t_k x_k$. The proof of Theorem 1 then applies with the omission of the last step.

3. *If* A *is coregular,* μ is uniquely determined by (1), by 4.4.8, and *the converse of Theorem 1 holds: if* $\mu = 0$ *no such* B *exists*. This is because $\chi(B) = \chi(f) = \mu\chi(A)$ (4.4.7) and B is coregular by 3.5.4.

4. In case A is not μ-unique the result of Theorem 1 holds for arbitrary f since a representation may be chosen with $\mu \neq 0$. The remaining case is covered in 15.5.3.

5.2. REPLACEABILITY

A matrix A is called *replaceable* if there exists a matrix B with null columns and $c_B = c_A$. If A is conull, B must be multiplicative-0 of course (3.5.4) and a coregular matrix A is replaceable iff a regular matrix B exists with $c_B = c_A$.

1. THEOREM. *Let* A *be coregular. Then the following are equivalent: (i)* A *is replaceable, (ii)* $1 \notin \overline{c_0}$ *(closure in* c_A*), (iii)* $1 \notin \overline{\phi}$ *(closure in* c_A*), (iv)* lim *is continuous on* c *as a*

subspace of c_A, *(v)* $x \to ax$ *is continuous on* c *as a subspace of* c_A.

(ii) = (iii) by 4.2.7 with $Y = c_A$, $X = c_o$, $E = \phi$.

(i) implies (iv): With $c_B = c_A$, B regular, we have lim $= \lim_B$ on c, so continuity is guaranteed by invariance of topology (4.2.4).

(iv) implies (i): Let $f \in c_A'$, f = lim on c by the Hahn-Banach theorem. Then $1 = \chi(f) = \mu(f)\chi(A)$ (4.4.7), so $\mu(f) \neq 0$. The result follows by 5.1.1.

(iv) implies (ii): Since lim = 0 on c_o and lim 1 = 1.

(ii) implies (iv): In c, with the c_A topology, c_o is a maximal subspace which is not dense. Hence it is closed (5.0.1). But $c_o = \lim^{\perp}$ so lim is continuous (5.0.1).

(iv) = (v): For $x \in c$, $\lim_A x = \chi(A)\lim x + ax$ (1.3.8). Since $\lim_A$ is continuous and $\chi(A) \neq 0$ the equivalence follows.

2. If A is conull, (ii), (iii), (iv) are always false and (v) is always true, while A may or may not be replaceable. To see this suppose f = 0 on c_o. Since $\chi(f) = \mu\cdot\chi(A) = 0$ it follows that f(1) = 0, so the falsity of (ii) follows by the Hahn-Banach theorem. Equivalence of this with (iv) is contained in the proof of Theorem 1. As to (v), for $x \in c$, $ax = \lim_A x$ by 1.3.8. Finally any multiplicative 0 matrix is automatically replaceable; a non-replaceable conull matrix is given in 13.2.7.

3. DEFINITION. *A conservative FK space is called regular if* $1 \notin \bar{c}_o$.

By 5.0.2 a regular space is coregular. By Theorem 1, a coregular matrix A is replaceable if and only if c_A is regular. It is not too difficult to construct a coregular space which is not regular — it takes more effort to construct one of the form c_A, equivalently a coregular non-replaceable matrix. This is done in Example 5.

4. THEOREM. *Let* A *be a conservative triangle whose columns are fundamental in* c. *Then* c_A *has* AD.

Let a^k be the kth column of A i.e. $a^k_n = a_{nk}$. Then $A^{-1}a^k = \delta^k$ and so $\{\delta^k\}$ is fundamental in c_A since $A^{-1}:c \to c_A$ is a linear isometry onto.

This result fails if A is not assumed to be a triangle; for example let A be the identity matrix with an extra column of 1's placed on its left side. Then $c_A = c$.

5. EXAMPLE. *A coregular perfect non-replaceable triangle. A coregular non-regular* BK *space of the form* c_A. AD *does not imply* AK *even for* $X = c_A$, A *a coregular triangle.* Let $b \in \ell$ with $b_k \neq 0$ for all k. Let $(Ax)_{2n-1} = x_{2n-1} + \sum_{k=1}^{n-1} b_k x_{2k}$, $(Ax)_{2n} = x_{2n-1} + \sum_{k=1}^{n} b_k x_{2k}$. The columns of A are $\delta^1+\delta^2$, $\delta^3+\delta^4$, $\delta^5+\delta^6,\ldots;$ $b_1(1-\delta^1)$, $b_2(1-\delta^1-\delta^2-\delta^3),\ldots$. The span of this set is the same as that of

$$\delta^1+\delta^2,\delta^3+\delta^4,\delta^5+\delta^6,\ldots;1-\delta^1,1-\delta^3,1-\delta^5,\ldots .$$

Now suppose that $f \in c'$, $f = 0$ on the columns of A. Then $0 = f(1-\delta^{2n-1}) = f(1) - f(\delta^{2n-1})$ hence $f(1) = 0$ since $f(\delta^n) \to 0$ (1.0.2). It follows that $f(\delta^{2n-1}) = 0$ for each n. But also $f(\delta^n+\delta^{n+1}) = 0$ for each n and so $f(\delta^n) = 0$ for all n. Thus finally $f = 0$. This shows, by the Hahn-Banach theorem that the columns of A are fundamental in c. By Theorem 4, c_A has AD hence Theorem 1 (iii) is satisfied and so A is not replaceable. It is easy to check that A is coregular. Of course AD implies perfect. Finally c_A does not have AK since $1 \in c_A$ and A is coregular.

6. THEOREM. *A non-μ-unique matrix (4.4.4) must be replaceable.*

There is a representation of the function 0 with $\mu \neq 0$ so by 5.1.1, there exists B with $c_B = c_A$, $\lim_B = 0$.

7. EXAMPLE. With A as in 4.4.4, one sees directly that $0 = \lim_A x - ax$. The preceding theorem may be cited or it may simply be checked that $(Bx)_n = \sum_{k=n}^{\infty} a_k x_k$ does the job. (Note that B is not a triangle. It is severely limited in this direction — see 15.2.14).

We continue with a few remarks to illuminate the conditions of Theorem 1.

8. DEFINITION. *The linear span of* ϕ *and* 1 *is denoted by* ϕ_1. *It is the set of eventually constant sequences.*

9. THEOREM. *Let* A *satisfy* $c_A \supset \phi_1$. *Then* $1 = \Sigma\, \delta^k$ *in* c_A *iff* $\Sigma_k\, a_{nk}$ *is uniformly convergent.*

First observe that the equality holds iff $\|1-1^{(m)}\|_A \to 0$ (4.3.13) because this condition is automatically satisfied by the other seminorms of c_A by 4.3.8. This expression is $\sup_n |\sum_{k=m+1}^{\infty} a_{nk}|$.

10. COROLLARY. *Let* A, B *be equipotent matrices with* $c_A \supset \phi_1$. *Then if* $\Sigma_k\, a_{nk}$ *is uniformly convergent, so is* $\Sigma\, b_{nk}$.

This is an invariance result and Theorem 9 "names" the property. (Naming program.)

11. DEFINITION. *A conservative FK space is called strongly conull if* $1 = \Sigma\, \delta^k$, *a matrix* A *is called strongly conull if* c_A *is strongly conull.*

Theorem 9 characterizes the spaces among conservative convergence domains.

12. EXAMPLE. *A coercive matrix is strongly conull* (1.7.18, Theorem 9), but *not conversely* e.g. let $a_{nk} = (-1)^k/[k \log(n+1)]$.

13. THEOREM. *Each of the following properties of a conservative FK space X implies the next but is not implied by it even if $X = c_A$, A a triangle. (i) AK, (ii) strongly conull, (iii) conull, (iv) $1 \in \overline{\phi}$.*

(iii) implies (iv): If $f = 0$ on ϕ, $f(1) = \chi(f) = 0$ and the result follows by the Hahn-Banach theorem. That (iv) does not imply (iii) is Example 5 with Theorem 1. That (iii) does not imply (ii) is shown by taking $(Ax)_n = x_n - x_{n-1}$ and applying Theorem 9. That (ii) does not imply (i) is shown by taking $(Ax)_n = x_n/n$. It obeys (ii) by Theorem 9; it is multiplicative 0 hence not perfect (3.3.5) and so by (iv) it does not even have AD.

If coregular is omitted in Example 5 an example can be given with easier calculations. (Note, X^f is defined in 7.2.3).

14. EXAMPLE. *A BK space X with AD and not AK, and $X^f \neq X^\beta$.* Let $X = c_A^0$ where $(Ax)_n = x_n - x_{n-1}$ $(x_0 = 0)$. If $f \in X'$, $f = 0$ on ϕ, then $f(x) = t(Ax)$ (4.4.2) and $0 = f(\delta^k) = (tA)_k$ so that $t \perp A$. This easily implies that $t = 0$ (A is of Type M) and AD follows by the Hahn-Banach theorem (3.0.1). Now, by Theorem 9, $1 \neq \Sigma \delta^k$ in c_A, hence in X by 4.3.14 — note that $1 \in X$. Thus X does not have AK. It turns out (10.6.3) that this is enough to imply $X^f \neq X^\beta$; with our present state of knowledge this requires a construction: let $y_n = \log(n+1)$, choose $t \in \ell$ so that $\lim t_n \log n$ does not exist and set $f(x) = t(Ax)$, $u_k = f(\delta^k)$. Then $u \in X^f \setminus X^\beta$ since $\sum_{k=1}^{n} u_k y_k = \sum_{k=1}^{n} t_k(Ay)_k - t_{n+1}y_n$ by 1.2.9.

5.3. CONSISTENCY

This subject is continued from Section 3.3 in which it was treated for coregular triangles. We first see that Type M drops out of the picture for general coregular matrices. That a perfect coregular matrix need not be of type M is shown trivially by taking A to be the identity matrix except that $a_{11} = 0$. The failure of the converse is shown in the next example.

1. EXAMPLE. If A is a *superdiagonal* matrix i.e. $a_{nk} = 0$ for $k < n$, and if $a_{nn} \neq 0$ for all n, A is trivially of Type M. However A need not be perfect. For example let A be the Tauberian matrix given in 1.8.6. It follows from 4.5.3 that c is closed. Of course A is coregular; indeed a multiple of a regular matrix.

2. THEOREM. *Let $A \in \Gamma$. The following conditions are equivalent (i) A is consistent with every stronger matrix B such that $\lim_B = \lim_A$ on c; (ii) A is consistent with every equipotent matrix B such that $\lim_B = \lim_A$ on c; (iii) A is perfect.*

(iii) implies (i): This follows from the continuity of $\lim_A$ and $\lim_B$ on c_A.

(ii) implies (iii): If A is not perfect, the Hahn-Banach theorem supplies $g \in c_A'$ with $g = 0$ on c, $g \neq 0$. We can also arrange that $\mu \neq 1$ in some representation of g. This is automatic if A is not μ-unique; otherwise multiply g by 2 if necessary. Let $f = \lim_A - g$. Then $\mu \neq 0$ in some representation of f and so, by 5.1.1, there exists B equipotent with A, $\lim_B = f$. This shows that (ii) is false.

3. THEOREM. *Let A be a coregular matrix. Then every stronger matrix B such that $\lim_B = \lim_A$ on c is consistent with A on $c_A \cap \ell^\infty$.*

This is immediate from 4.6.7.

Theorem 2 is more general than 3.3.7 in that the latter is restricted to coregular matrices — but less specific in that 3.3.7 gives an internal test (Type M). A remedy for this is suggested in the introduction of P, Chapter 15.

5.4. REVERSIBLE MATRICES

1. DEFINITION. *A matrix* A *is called reversible if for each* $y \in c$ *there is a unique* x *such that* $Ax = y$.

For example, a triangle is reversible. A matrix may have a (two-sided) inverse without being reversible, for example the matrix A in 1.8.6 is not one to one.

2. EXAMPLE. Let u, v be sequences with $u_n \to 1$, $v_n \neq 0$ for all n. (For a first reading take $u = v = 1$.) Let $(Ax)_n = u_n x_1 + \sum_{k=n+1}^{\infty} v_k x_k$; A is one to one. Also, if $y \in c$, let $x_1 = \lim y$, $x_n = [(u_n - u_{n-1})/v_n]\lim y + (y_{n-1} - y_n)/v_n$ for $n > 1$. (Convention: $u_0 = 0$). Then $Ax = y$ so A is reversible.

3. THEOREM. *Let* A *be reversible. Then* c_A *is a* BK *space with* $\|\cdot\|_A$ *and the general form of* $f \in c_A'$ *is* $f(x) = \mu \lim_A x + t(Ax)$, $t \in \ell$.

As in §3.2 it is seen that c_A is a Banach space with the dual representation shown. The fact that coordinates are continuous lies much deeper. It follows from 4.3.2.

4. If f is represented as in Theorem 3, μ is uniquely determined. However f will always have representations as in 4.4.3 with $\alpha \neq 0$ and sometimes with different values of μ, even if A is a triangle! (4.4.4)

Reversible matrices first rose in the problem of solving infinite systems of linear equations. The existence and form of the solution are shown as follows:

5. THEOREM. *Let* A *be a reversible matrix. Then* A *has a unique right inverse* B. *The rows of* B *belong to* ℓ. *There is a sequence* b *such that the equation* $y = Ax$ *has, for* $y \in c$, *the unique solution* $x = b \lim y + By$.

Applying Theorem 3 to the coordinates we have, for $x \in c_A$ and $y = Ax$, $x_n = \mu_n \lim_A x + t_n(Ax) = \mu_n \lim y + \Sigma\, t_{nk}y_k$. Now setting $b_n = \mu_n$, $B = (t_{nk})$ we have all the theorem except that B is a right inverse. We see this by taking $y = \delta^k$; then $x_n = (B\delta^k)_n = b_{nk}$ (i.e. t_{nk}). The equation $y = Ax$ becomes $\delta^k_n = \Sigma_i\, a_{ni}x_i = \Sigma_i\, a_{ni}b_{ik}$ i.e. $AB = I$.

Example 2 shows that b may be unbounded. (This was first shown by M.S. MacPhail.) The next result shows the futility of trying to modify Example 2 so as to make A regular or row finite:

6. THEOREM. *Let* A *be a reversible matrix which is either* μ-*unique (e.g. coregular by 4.4.8), or row-finite. Then* $b = 0$ *in Theorem 5.*

If A is μ-unique, the numbers μ_n occuring in the proof of Theorem 5 are 0 since x_n has $\mu = t = 0$, $\alpha = \delta^n$ in 4.4.3. If A is row-finite, let $y = 1 = Ax$, then from Theorem 5, $1 = (Ax)_n = (Ab)_n + [A(B1)]_n = (Ab)_n + [(AB)1]_n$ by 1.4.4. This is $(Ab)_n + 1$ and so $Ab = 0$. Hence $b = 0$.

7. THEOREM. *Let* A *be a reversible matrix with convergent columns and* B, b *as in Theorem 5. Then* $BA = I - D$ *where* $d_{nk} = b_n a_k$, $a_k = \lim_n a_{nk}$.

From Theorem 5, $x = b \lim Ax + B(Ax)$ for each $x \in c_A$. Taking $x = \delta^k$ gives the result.

8. It follows that $B = A^{-1}$ if either $b = 0$ or A has null columns. If $b = 0$ A^{-1} gives the inverse transformation by Theorem 5, but this is not so if A has null columns. In [83], Theorem 11, there is given a multiplicative 0 reversible matrix (hence $B = A^{-1}$ as above) with $b \neq 0$ and so A^{-1} does not give the inverse transformation.

9. If A is coregular or row-finite with convergent columns, $B = A^{-1}$. These follow from Theorems 6, 7. With perserverence one can squeeze out a little more: if A is row-finite with bounded columns, BA exists and $A(BA) = (AB)A = A$ (1.4.4) so $B = A^{-1}$.

10. EXAMPLE. The simplest case, $u = v = 1$, of Example 2 shows a reversible matrix with no left inverse.

11. THEOREM. *Let A be coregular and reversible. Then A is perfect iff it is of type M.*

If $f = 0$ on c we have $f(x) = t(Ax)$ with $t \in \ell$, $tA = 0$. This follows from Theorem 3 and 4.4.7 along with $f(\delta^k) = 0$ for all k. So if A is of Type M, $f = 0$ and A is perfect, by the Hahn-Banach theorem. Conversely, let $tA = 0$, $t \in \ell$. Let $f(x) = t(Ax)$. Then $f = 0$ on c (1.4.4) hence $f = 0$. Fix n and choose x such that $Ax = \delta^n$. Then $0 = f(x) = t_n$ i.e. $t = 0$.

With 5.3.2 this gives a consistency theorem identical with that proved for triangles. (3.3.7)

12. REMARK (M.S. MacPhail). We can now point out the lack of significance of weak μ-uniqueness. Call A *weakly μ-unique* if μ is uniquely determined by f in 4.4.3 with $\alpha \in \omega_A^\beta$. A *row-finite reversible matrix* need not be μ-unique even if it is a triangle, (4.4.4), but *must be weakly μ-unique*. [If not there

exist $t \in \ell$, $\alpha \in \phi$ with $\lim_A x + t(Ax) + \alpha x = 0$. Each $x_i = [B(Ax)]_i$ by Theorems 5, 6 so, setting $y = Ax$, we have $x_i = \sum_{k=1}^{\infty} b_{ik}y_k$ and $\alpha x = \sum_{k=1}^{\infty} u_k y_k$ where $u_k = \sum_{i=1}^{m} \alpha_i b_{ik}$. Hence for all $y \in c$, $0 = \lim y + ty + uy$. Setting $y = \delta^k$ gives $t + u = 0$. Then $y = 1$ yields a contradiction.] (Example 2 shows that this fails without row-finiteness.) It follows that *weak* μ*-uniqueness is not invariant* by considering 5.2.7. There A is a triangle, $c_A = c_B$, and $\lim_B = 0$ so B is not weakly μ-unique.

5.5. ROW-FINITE AND ONE TO ONE

1. THEOREM. *Let A be a row-finite matrix. Then $A[\omega]$ is a closed subset of ω.*

First, A is continuous by 4.2.8. The result is that given in 5.0.3. An elementary proof is given in [79] #6.4.28.

2. COROLLARY. *If A is row-finite and reversible the equation $y = Ax$ can be solved for every y.*

For $A[\omega] \supset c$ which is dense in ω.

Row-finite cannot be dropped. In 5.4.2 if $y = Ax$ it follows that $y_n \to x_1$ i.e. $y \in c$.

Row-finite one to one matrices behave like reversible matrices in that they obey analogoues of 5.4.3 and 5.4.11.

3. THEOREM. *Let A be row-finite and one to one. Then c_A is a BK space with $\|\cdot\|_A$ and every $f \in c_A'$ can be written $f(x) = \mu \lim_A x + t(Ax)$, $t \in \ell$.*

Let $Y = c \cap A[\omega]$ then by Theorem 1 and 4.2.15 Y is an FK space with $(\|\cdot\|_\infty, p)$ where $p_n(x) = |x_n|$. Since $|x_n| \le \|x\|_\infty$ all the seminorms p are redundant so Y is a BK space with $\|\cdot\|_\infty$. This could also be seen by checking directly that (Y, T_ω)

is closed in c, hence *a fortiori* (Y, T_c) is closed in c. We now have that $A : c_A \to Y$ is one to one and onto and 4.3.2 yields the first part. (Take $X = \omega$, $A^{-1}[Y] = c_A = Z$). If $f \in c_A'$, let $g(y) = f(x)$ for $y \in Y$, $y = Ax$. Extending g to all of c by the Hahn-Banach theorem gives $f(x) = g(y) = \mu \lim y + ty = \mu \lim_A x + t(Ax)$.

4. THEOREM. *A row-finite one to one coregular matrix is perfect if it is of type M. The converse is false.*

This is proved in the same way as 5.4.11. For the converse, modify the identity matrix by adding a row of zeros on top or repeating each row or some such.

5. THEOREM (H. SKERRY). *Let* A *be a row-finite matrix. Then* A *has a growth sequence iff* $A^\perp = \{x : Ax = 0\}$ *has finite dimension.*

Necessity: If $A^\perp$ has infinite dimension it has no growth sequence by 4.2.11, 4.2.12 and the fact that $A : \omega \to \omega$ is continuous (4.2.8). The result follows since $c_A \supset A^\perp$.

Sufficiency: We may assume A is one to one since placing finitely many finite rows on top of A does not alter c_A. Then c_A is a BK space by Theorem 3 and the result follows by 4.2.11.

6. It is not known whether sufficiency holds in general. Necessity fails since, for example, if $a_{1k} = 1$, $a_{nk} = 0$ for $n > 1$ $c_A = cs$ which has 1 as a growth sequence. Of course a matrix must have a growth sequence if it has a row with no zeros in it.

5.6. BOUNDED CONSISTENCY

Various subspace of an FK space, called distinguished, will be studied systematically beginning in Chapter 10. One of them is the space W which it will be useful to define now.

1. DEFINITION. *Let* X *be an* FK *space* $\supset \phi$. *Then* $W = W(X) = \{x \in X: x = \Sigma x_k\delta^k$, *convergence in the weak topology of* $X\}$. *Also* $W_b = W \cap \ell^\infty$. *If* A *is a matrix* $W(A) = W(c_A)$.

Equivalently: $x \in W$ iff $f(x) = \Sigma x_k f(\delta^k)$ for all $f \in X'$; $x \in W$ iff $x^{(n)} \to x$ weakly, where $x^{(n)}$ is, as usual, the nth section of x. It is customary to write $x \in W$ as x has SAK, standing for schwach AK (= weak AK).

2. A conservative space is conull iff $1 \in W$.

3. DEFINITION. *Let* A *be a matrix with convergent columns,* $a_k = \lim a_{nk}$. *Then* $\Lambda(x) = \lim_A x - ax$ *for* $x \in a^\beta \cap c_A$.

4. The definition has been given in more generality than is required now. Its ramifications will be explored beginning in §13.2. For now *it will be assumed that* $A \in \Gamma$ *in which case* Λ *is defined on* $c_A \cap \ell^\infty$; for $a \in \ell$ (1.3.7) and so $a^\beta \supset \ell^\infty$.

5. In particular $\Lambda(1) = \chi(A)$ so A is conull iff $\Lambda(1) = 0$. (With the assumption of Remark 4.)

6. LEMMA. *Let* $A \in \Gamma$, $x \in c_A \cap \ell^\infty$. *Then* $x \in W$ *iff* $\Lambda(x) = 0$.

If $x \in W$, apply the definition of W to $f = \lim_A$; thus $\lim_A x = \Sigma x_k \lim_A \delta^k = ax$. Conversely if $\Lambda(x) = 0$, let $f \in c_A'$. Then by 4.4.9, $f(x) = \mu \lim_A x + \gamma x = \mu ax + \gamma x$ by hypothesis. Also $\Sigma x_k f(\delta^k) = \Sigma x_k(\mu a_k + \gamma_k) = f(x)$, hence $x \in W$.

Attachment was introduced in 4.3.4; for matrices it takes this form:

7. DEFINITION. *Let* z *be a sequence,* A *a matrix. The matrix* $A \cdot z$ *is* $(a_{nk}z_k)$.

Thus $Y_{A\cdot z} = z^{-1}\cdot Y_A$ for any space Y.

8. LEMMA. *Let* $A \in \Gamma$, $z \in c_A \cap \ell^\infty$. *Then* $z \in W$ *if and only if* $A\cdot z$ *is conull.*

This follows from the fact that $\chi(A\cdot z) = \Lambda(z)$, with Lemma 6. This result is given a more natural setting in 12.1.6.

9. LEMMA. *Let* $A,B \in \Gamma$ *with* $c_A \cap \ell^\infty \subset c_B$. *Then* $W_b(A) \subset W(B)$.

If $z \in W_b(A)$, $A\cdot z$ is conull by Lemma 8 and so $B\cdot z$ is conull by 3.5.4. Hence $z \in W(B)$ by Lemma 8.

We can now give a proof of the famous *Mazur-Orlicz Bounded Consistency Theorem.* About 5 proofs of this theorem are in the literature. The simplest one, given here, is from [70]. The result strengthens 3.3.6 (ii), which has the same conclusion in that its hypothesis is weaker.

10. THEOREM. *Let* A, B *be regular matrices with* $c_A \cap \ell^\infty \subset c_B$. *Then* A, B *are consistent for bounded sequences.*

We prove instead a more general result which similarly improves 3.3.7 (ii).

11. THEOREM. *Let* A, B *be coregular matrices with* $c_A \cap \ell^\infty \subset c_B$ *and* $\lim_A x = \lim_B x$ *for* $x \in c$. *Then* A, B *are consistent for bounded sequences.*

Note first that $b_k = \lim_B \delta^k = \lim_A \delta^k = a_k$. Let $z \in c_A \cap \ell^\infty$.

CASE I: $\Lambda_A(z) = 0$. Then $\Lambda_B(z) = 0$ by Lemmas 9 and 6. Thus $\lim_B z = bz$ and $\lim_A z = az$. These are equal as just mentioned.

CASE II: Let $y = z - t1$ where $t = \Lambda_A(z)/\Lambda_A(1)$. (See Remark 5). Then $\Lambda_A(y) = 0$ so $\lim_B y = \lim_A y$ by Case I. Since $\lim_B 1 = \lim_A 1$ it follows that $\lim_B z = \lim_A z$.

The failure of this result for unbounded sequences is demonstrated in [26].

CHAPTER 6
BIGNESS THEOREMS

6.0. FUNCTIONAL ANALYSIS

1. A reflexive BK space X must be separable: Let $P_n(x) = x_n$, then $\{P_n\}$ is total over X hence fundamental in (X', W^*) [80], Theorem 8-1-10; hence the latter space is separable. Since the weak * and weak topologies coincide X' is weakly separable, hence norm separable [80], #8-3-103; so X is separable, [80], #9-5-4.

2. Let X be a separable BK space, D the unit disc in X'. Then (D, W^*) is metrizable and compact, hence sequentially compact. [80], Theorem 9-1-12, #9-5-1.

3. The metric for an FK space (X,p) is given by $d(x,y) = \|x-y\|$ where $\|x\| = \Sigma\, 2^{-n}p_n(x)/[1+p_n(x)]$. [80], Theorem 2-1-2. We shall prove that $\|tx\| \leq \|x\|$ if $|t| \leq 1$: if p is a seminorm $p(tx)/[1+p(tx)] = (1+1/[|t|p(x)])^{-1} \leq [1+1/p(x)]^{-1} = p(x)/[1+p(x)]$. Multiplying by 2^{-n} and summing gives the result.

4. Let X be a Fréchet space with topology given by a set P of seminorms. If $\Sigma\, p(x^n) < \infty$ for each $p \in P$, then $\Sigma\, x^n$ is convergent: for if U is a neighborhood of 0, $U \supset \cap\{x: p(x) < \varepsilon\}$ where the intersection is taken over a finite subset F of P. [80], Theorem 4.1.12. With $y^m = \sum_{n=1}^{m} x^n$ we have for $p \in F$, $p(y^m - y^k) \leq \sum_{k+1}^{m} p(x^n) < \varepsilon$ for large k, m. Thus, for large k, m,

$y^m - y^k \in U$ and so $\{y^m\}$ is a Cauchy sequence, hence convergent.

6.1. c NOT CLOSED

The two main results of this chapter are that if c is not closed in a conservative FK space X, then X contains bounded divergent sequences (6.1.1) and that if X is conull (in which case c is not closed by 4.6.3), then X must include a large space of a certain specified form (6.4.2). The first result was proved earlier under rather restrictive hypotheses (3.4.4); the extension is an item in the FK program. Recall also that if c is not closed, X has lots of unbounded sequences too (4.5.6).

1. THEOREM. (W. Meyer-König and K. Zeller). *Let X be a conservative FK space in which c is not closed. Then X contains bounded divergent sequences.*

We may assume that the topology X is given by a sequence q of seminorms satisfying $q_{n+1} \geq q_n$ for all n, for if X is a BK space we may take all q_n to be the same, otherwise replace q_n by $\sum_{k=1}^{n} q_k$.

Fix an integer $u > 1$ and set $Z = \{x \in c_o: x_1 = \ldots = x_{u-1} = 0\}$. Then Z has codimension u in c, hence is not closed in X by 3.0.3. So on Z the $\|\cdot\|_\infty$ topology is strictly larger than the relative topology of X (4.2.4) hence strictly larger than that induced by any one q_i i.e. for any $\varepsilon > 0$ and any i there exists $x \in Z$ such that $\|x\|_\infty = 1$, $q_i(x) < \varepsilon$. To summarize: Given $\varepsilon > 0$, integer $u > 1$, integer i, there exist x and integer $v > u$ such that

$$x_k = 0 \text{ for } 1 \leq k < u, \ |x_k| < \varepsilon \text{ for } k > v, \ \|x\|_\infty = 1 \qquad (1)$$

and

$$x \in X, \ q_i(x) < \epsilon \tag{2}$$

The existence of v is guaranteed since $x \in c_o$.

Choose x^1, v_1 to satisfy (1), (2) with $\varepsilon = 1/8$, $u_1 = 2$, $v_1 > u_1$, $i = 1$; then x^2, v_2 with $\varepsilon = 1/16$, $u_2 = v_1 + 2$, $v_2 > u_2$, $i = 2; \ldots; x^n, v_n$ with $\varepsilon = 2^{-n-2}$, $u_n = v_{n-1} + 2$, $v_n > u_n$, $i = n$. Now let $x = \Sigma x^n \in X$. The series converges since for each i, $q_i(x^n) \leq q_n(x^n) < 2^{-n-2}$ as soon as $n \geq i$ and so $\Sigma q_i(x^n)$ is convergent. (6.0.4). Also $x_k = \Sigma_n x_k^n$ since X is an FK space. We have arranged the notation so that the check that x is bounded and divergent is simply the last part of the proof of 3.4.4, the line (1) being the same in both places, as well as the choice of x^n, v_n; the extra condition $(i=n)$ being irrelevant.

We can now incorporate 3.5.4 into the FK program but only partially. A complete extension would be: if Y is conull and $X \supset Y \cap \ell^\infty$ then X is conull; but this is false as shown by $Y = \omega$, $X = \ell^\infty$. In spite of this a complete extension will be given in 16.2.7.

2. THEOREM. *Let X be conull and A a matrix with $c_A \supset X \cap \ell^\infty$. Then A is conull.*

Let $B = A - \chi(A)I$ so that B is conull. Hence $c_B \cap X$ is conull (4.6.6) and so, by Theorem 1, contains a bounded divergent x. By hypothesis $x \in c_A$, then $\chi(A)x = Ax - Bx \in c$ and so $\chi(A) = 0$.

3. THEOREM. *Let $A \in \Gamma$. Then A is Tauberian iff c is closed in c_A.*

Necessity is by Theorem 1. Conversely, suppose that c is closed. Then A is coregular (4.6.4) and the result follows by 4.6.7.

4. Theorem 3 was given for triangles in 3.4.4. The FK program fails; for example, let $X = c \oplus z$ where z is bounded and divergent (4.5.5).

The next result improves 1.8.4:

5. COROLLARY (J. Copping). *Let A be a conservative matrix which has a left inverse B of finite norm. Then A is Tauberian. If B is also row-finite, A is Mercerian.*

For $x \in c$, $\|x\|_\infty = \|(BA)x\|_\infty = \|B(Ax)\|_\infty$ (1.4.4) $\leq \|B\| \cdot \|x\|_A$ $\leq \|B\| \cdot \|A\| \cdot \|x\|_\infty$. Thus c is closed in c_A (4.2.4) and A is Tauberian, by Theorem 3. If B is row-finite the same argument may be applied to every $x \in c_A$ proving that x is bounded. The result follows by 4.6.8.

This result is placed in another context and given a converse in 18.1.7, 18.1.10.

6. COROLLARY. *Suppose that $A \in \Gamma$ has a two-sided inverse B with $\|B\| < \infty$.* Then $B \in \Gamma$.

Let $x \in c$. Then $Bx \in c_A \cap \ell^\infty$ since $A(Bx) = x \in c$ (1.4.4) and $\|Bx\|_\infty \leq \|B\| \cdot \|x\|_\infty$. So $Bx \in c$ by Corollary 5.

(I made a nice little problem out of this — it appears as #6414 in the December 1982 American Mathematical Monthly.)

7. LEMMA. *Let $A, B \in \Gamma$ with B conull. Then AB is conull.*

If x is bounded, $(AB)x = A(Bx)$ (1.4.4), so $c_{AB} \supset c_B \cap \ell^\infty$. The result follows by Theorem 2.

8. COROLLARY. *For $A, B \in \Gamma$, $\chi(AB) = \chi(A)\chi(B)$.*

Let $M = B - \chi(B)I$. Then AM is conull by Lemma 7 so $0 = \chi(AM) = \chi(AB) - \chi(B)\chi(A)$.

9. COROLLARY. *Let* $A, B \in \Gamma$ *with* B *conull. Then* AB *and* BA *are conull.*

By Corollary 8.

SOURCES: [14], [48A], [83].

6.2. TWO-NORM CONVERGENCE AND W.

The concept of two-norm convergence will be introduced with more generality than necessary. For purpose of this chapter the reader may take $p = \|\cdot\|_\infty$ in the following definition. The phrases "$p(x) < \infty$" and "$p(x)$ is defined" are synonymous.

1. DEFINITION. *Let* X *be an* FK *space* $\supset \phi$ *and* p *a norm defined on a subspace of* X. *Then* p *is called admissible for* X *if* $\{x: p(x) \leq 1\}$ *is closed in* X *and, for all* k, $p(x^{(k)}) \leq p(x)$ *whenever* $p(x) < \infty$.

Recall that $x^{(k)}$ is the kth section of x. (4.2.13).

2. EXAMPLE. *Let* X *be an* FK *space, then* $\|\cdot\|_\infty$ *is admissible for* X. Let $D = \{x \in X: \|x\|_\infty \leq 1\}$. Then $D = \cap D_n$ where $D_n = \{x \in X: |x_n| \leq 1\}$. Each D_n is closed in X since X is an FK space.

3. DEFINITION. *Let* X *be an* FK *space* $\supset \phi$ *and* p *an admissible norm. Let* $E \subset X$. *Then* 2_pE *is defined to be* $\{x \in X$*: there exists a sequence* $\{e^n\}$ *of points of* E *with* $e^n \to x$ *in* X *and* $\{p(e^n)\}$ *bounded*$\}$. *For* $p = \|\cdot\|_\infty$ *we write* 2_pE *as* $2_\infty E$.

The use of 2 is to indicate the presence of two topologies. Thus 2_pE is called the two-norm closure of E (even though X need not be a normed space.) It may be a good deal smaller than the ordinary closure, for example take $X = \omega$ in Theorem 6. It may even be empty! (Take $E \not\subset Z$ in Lemma 4).

4. LEMMA. *With X, p, E as in Definition 3, $2_p E \subset Z$ where $Z = \{x \in X: p(x) < \infty\}$.*

Let $z \in 2_p E$, say $e^n \to z$ with $p(e^n) \leq M$. Then $p(z) \leq M$ since the disc $(p \leq M)$ is closed in X.

5. LEMMA. *Let X be an FK space $\supset \phi$ and p an admissible norm. Then $W \cap Z \subset 2_p \phi$. (For W see 5.6.1).*

Here, as before, $Z = \{x \in X: p(x) < \infty\}$. Let $z \in W \cap Z$. By definition $z^{(n)} \to z$ weakly, so z lies in the closure of the convex hull of $\{z^{(n)}\}$. (5.0.2). Say $a^n \to z$ in X where $a^n = \Sigma\{t_i^n z^{(i)}: 1 \leq i \leq m(n)\}$, $t_i^n \geq 0$, $\Sigma_i t_i^n = 1$ for each n. Then $p(a^n) \leq \Sigma\, t_i^n\, p(z^{(i)}) \leq p(z)\, \Sigma\, t_i^n = p(z)$. Each a^n clearly lies in ϕ.

6. THEOREM. *Let X be a conservative FK space. Then $W_b = 2_\infty \phi = 2_\infty c_0$.*

Here $W_b = W \cap \ell^\infty$. Let $z \in 2_\infty c_0$. In view of Lemmas 4, 5 and Example 2 it is sufficient to prove that $z \in W$. To this end let $f \in X'$. Say $a^n \to z$, $a^n \in c_0$, $\|a^n\|_\infty < M$. Then

$$f(z) = \lim f(a^n) = \lim_n \Sigma_k\, a_k^n f(\delta^k) \tag{1}$$

since $a \in c_0$ implies $a = \Sigma\, a_k \delta^k$ in c_0, *a fortiori*, in X. Now $|a_k^n f(\delta^k)| \leq M|f(\delta^k)|$ and $\Sigma|f(\delta^k)| < \infty$ since $X \supset c_0$ (1.0.2, 4.2.4). Hence the series in (1) is uniformly convergent and so we may let $n \to \infty$ inside the summation. This leads to $f(z) = \Sigma\, z_k f(\delta^k)$ since $a^n \to z$ and X is an FK space. This proves that $z \in W$.

7. COROLLARY (A.K. Snyder). *A conservative FK space is conull if and only if $1 \in 2_\infty \phi$.*

This is by Theorem 6 and 5.6.2.

8. The condition $1 \in \overline{\phi}$ is weaker than that of Corollary 7 and does not imply conull (5.2.1,5.2.5).

Applications and historical discussion of two-norm topologies are given in [12], 564-565.

6.3. OSCILLATION

1. DEFINITION. *Let* r *be a strictly increasing sequence of positive integers with* $r_1 = 1$. *Then* $O^r(X)$, *which we write* $O(X)$ *when* r *is fixed, is the sequence whose nth term*
$O_n(x) = max\{|x_u - x_v| : r_n \leq u < v \leq r_{n+1}\}$ *and*
$\Omega(r) = \{x : O_n(x) \to 0$ *as* $n \to \infty\}$; $\Omega_b(r) = \Omega(r) \cap \ell^\infty$.

Properties of Ω are most conveniently derived from a matrix representation. In the sequel $c_A^0 = \{x : Ax \in c_0\}$ as in 1.2.2.

2. THEOREM. *There exists a multiplicative 0 triangle A such that* $c_A^0 = \Omega(r)$.

Of course A depends on r. Let $(Ax)_1 = x_1$ and, for $n > 1$, $(Ax)_n = x_n - x_{r_i}$ where i is chosen so that $r_i < n \leq r_{i+1}$. Each row of A has two non-zero members, ± 1, thus the row sums are zero and $\|A\| = 2$. Further $r_i \to \infty$ as $n \to \infty$ since $r_{i+1} \geq n$ so each column of A terminates in zeros.

If $x \in \Omega(r)$, $|(Ax)_n| = |x_n - x_{r_i}| \leq O_i(x) \to 0$ so $x \in c_A^0$. Conversely, let $x \in c_A^0$, $\varepsilon > 0$. Choose j such that $n > r_j$ implies $|(Ax)_n| < \varepsilon/2$. Now let $i > j$ and $r_i \leq u < v \leq r_{i+1}$. If $u = r_i$, $|x_u - x_v| = |(Ax)_v| < \varepsilon/2$ while if $u > r_i$, $|x_u - x_v| \leq |x_u - x_{r_i}| + |x_{r_i} - x_v| = |(Ax)_u| + |(Ax)_v| < \varepsilon$. This proves that $O_i(x) < \varepsilon$. Hence $x \in \Omega(r)$.

3. COROLLARY. *Each* $\Omega(r)$ *is a conull conservative BK space.*

In Theorem 2, c_A is conull, c_A^0 includes ϕ and 1 and is a closed subspace of c_A since it is $\lim_A^{\perp}$. The result follows by 4.6.3.

4. COROLLARY. *Each* $\Omega_b(r)$ *is a closed subspace of* ℓ^∞.

By Theorem 2 and 1.3.11.

5. LEMMA. *Suppose that* r *is given. Then for each* $y \in \ell^\infty$ *there exists* $x \in \Omega_b(r)$ *such that* $x_{r_{2^n}} = y_n$.

It remains to complete the definition of x starting from the part given in the statement. Let x_{r_k} go in equal steps from y_r to y_{r+1} as k goes from 2^n to 2^{n+1}; finally let $x_i = x_{r_k}$ for $r_k \le i < r_{k+1}$. Then, for $2^n \le k < 2^{n+1}$, $0_k(x) = |x_{r_{k+1}} - x_{r_k}| = |y_{n+1} - y_n|/2^n \le 2\|y\|_\infty/2^n \to 0$ so $x \in \Omega(r)$. Also x is bounded.

6. In Lemma 5, $\{2^n\}$ *can be replaced by any sequence* u *such that* $u(n+1) - u(n) \to \infty$. The same proof works with 2^n replaced by $u(n)$ except in its last two occurrences where it is replaced by $u(n+1) - u(n)$.

7. LEMMA. *Suppose that* r *is given and* v *is an increasing sequence of integers such that no interval* $[r_n, r_{n+1})$ *contains more than one* v_k. *Then there is a subsequence* w *of* v *with the property: for each* $y \in \ell^\infty$ *there exists* $x \in \Omega_b(r)$ *such that* $x_{w_n} = y_n$.

Say $v_k \in [r_{m(k)}, r_{m(k)+1})$ and $w_n = v_{i_n}$ so that $w_n \in [r_{m(i_n)}, r_{m(i_n)+1}) = [r_{u(n)}, r_{u(n)+1})$; w can be chosen so that $u(n+1) - u(n) \to \infty$. By Remark 6 there exists, for each $y \in \ell^\infty$, an $x \in \Omega(r)$ such that $x_{r_{u(n)}} = y_n$. The construction of Lemma 5, with $u(n)$ instead of 2^n, shows that $x_{w(n)} = y_n$ also.

Now attachment takes a third form:

8. DEFINITION. *Let* E *be a set of sequences and* z *a sequence. Then* $z\cdot E = \{z\cdot x : x \in E\}$. *The notation* $z\cdot x$ *was introduced in 4.3.4.*

9. THEOREM. *Let* $z \in \ell^\infty \backslash c_0$, $Z = z\cdot\Omega_b(r)$ *for some* r. *Then* Z *is a non-separable subspace of* ℓ^∞. *In particular this applies with* $z = 1$.

Let S be a non-countable set of sequences of 0's and 1's. Let $|z_{v(n)}| \geq \varepsilon > 0$; v can be chosen to satisfy the hypotheses of Lemma 7. Then with w as in Lemma 7, there exists, for each $s \in S$ an $s' \in \Omega_b(r)$ with $s'_{w(n)} = s_n$. Now if $s,t \in S$ with $s \neq t$ we have $|s_n - t_n| = 1$ for some n. Then $\|z\cdot s' - z\cdot t'\|_\infty \geq |z_{w(n)}|\cdot|s'_{w(n)} - t'_{w(n)}| = |z_{w(n)}| \geq \varepsilon$. Thus Z contains a non-countable set all of whose pairwise distances $\geq \varepsilon$.

10. Lemma 7 may be interpreted as saying that there is a regular matrix A which maps $\Omega_b(r)$ onto ℓ^∞. Namely, let $a_{n,w_n} = 1$, $a_{nk} = 0$ for $k \neq w_n$.

6.4. CONULL SPACES

In this section we give the bigness theorem of A. K. Snyder, Theorem 2, which gives a best possible inclusion theorem for conull FK spaces; namely that such a space must include some $\Omega(r)$; it is best possible in that each $\Omega(r)$ is itself conull by 6.3.3. Partial results, now superseded, were given in 4.6.5.

Theorem 2 differs from earlier bigness theorems in that it gives a sufficient condition for summability, namely, every sufficiently slowly oscillating sequence is summable. Such theorems are called *direct theorems*. They are opposite in spirit to Merceriang and Tauberian theorems.

1. LEMMA. *Let X be an FK space $\supset \phi_1$ (5.2.8) such that $1 \in 2_\infty \phi$ (6.2.3). Then there exists r such that $X + c_0 \supset \Omega(r)$.*

Let $a^n \to 1$ in X, $a^n \in \phi$, $\|a^n\|_\infty$ bounded. Let $b^n = 1 - a^n$ so that $b^n \to 0$ in X; $b_k^n \to 0$ as $n \to \infty$ for each k; for each n, $b_k^n = 1$ for sufficiently large k; $\|b^n\|_\infty < M$.

The second condition holds because coordinates are continuous. Let $r_1 = 1$ and choose p_1 so that $\|b^{p_1}\| < \frac{1}{2}$. Inductively suppose that $r_1, r_2, \ldots, r_{n-1}$ and $p_1, p_2, \ldots, p_{n-1}$ have been chosen. Choose $p_n > p_{n-1}$ so that $\|b^{p_n}\| < 2^{-n}$ and $|b_k^{p_n}| < 2^{-n}$ for $k = 1, 2, \ldots, r_{n-1}$ and choose $r_n > r_{n-1}$ so that $b_k^{p_n} = 1$ for $k \geq r_n$.

Now let $y \in \Omega(r)$ and we shall show that $y \in X + c_0$, completing the proof.

First let $z^n = b^{p_n}$, $t_n = y_{r_n} - y_{r_{n-1}}$ $(t_1 = y_1)$ and $x = \Sigma\, t_n z^n$. This series converges by 3.0.5 since $|t_n| < 0_{n-1}(y) \to 0$ and $\|z^n\| < 2^{-n}$, hence, by 6.0.3, $\|t_n z^n\| < 2^{-n}$ eventually. Thus $x \in X$ and the proof is concluded by showing that $y - x \in c_0$.

Now $|y_k - x_k| \leq |y_k - \sum_{n=1}^{m} t_n z_k^n| + |t_{m+1} z_k^{m+1}| + \sum_{n=m+2}^{\infty} t_n z_k^n|$ $= T_1 + T_2 + T_3$, say. In this inequality we choose m so that $r_m \leq k < r_{m+1}$. Then $T_1 = |y_k - \sum_{n=1}^{m} t_n| = |y_k - y_{r_m}| \leq 0_m(y)$; $T_2 \leq |t_{m+1}| \cdot \|z^{m+1}\|_\infty \leq M|t_{m+1}| \leq M\, 0_m(y)$; in T_3 we see that $n-1 \geq m+1$ so that $r_{n-1} \geq r_{m+1} > k$ hence $T_3 \leq \sum_{n=m+2}^{\infty} |t_n| 2^{-n}$ $\leq \|t\|_\infty 2^{-m-1}$. Thus $|y_k - x_k|$ is dominated by the sum of three terms each of which $\to 0$ as $m \to \infty$, hence as $k \to \infty$.

2. THEOREM. *Every conull FK space includes $\Omega(r)$ for some r.*

By 6.2.7, $1 \in 2_\infty \phi$ and so Lemma 1 applies.

3. COROLLARY (a). *A matrix* A *is conull if and only if* $c_A \supset \Omega_b(r)$ *for some* r. *(b) An* FK *space is conull if and only if it includes* $\Omega(r)$ *for some* r.

Part (a) follows from Theorem 3, 6.1.2 and 6.3.3. Part (b) follows from Theorem 3, 4.6.3 and 6.3.3.

4. EXAMPLE. Since ℓ^∞ is coregular and includes every $\Omega_b(r)$, Part (b) of Corollary 3 cannot be improved to resemble Part (a).

5. COROLLARY. *Let* X *be a conull* FK *space. Then* $X \cap \ell^\infty$ *is a non-separable subspace of* ℓ^∞.

This is by Theorem 2 and 6.3.9.

6.5. COREGULAR SPACES AND MATRICES.

The conull theory may be extended by using attachment.

1. THEOREM. *Let* X *be an* FK *space* $\supset c_o$, $z \in X \cap \ell^\infty$, $Y = z^{-1}\cdot X$ *(4.3.4). Then* Y *is a conservative* FK *space. The general* $f \in Y'$ *is given by* $f(y) = \alpha y + g(z\cdot y)$, $\alpha \in \phi$, $g \in X'$.

That Y is an FK space is 4.3.6. If $u \in c_o$ then $u\cdot z \in c_o \subset X$, so $u \in Y$. Also $1\cdot z = z \in X$ so $1 \in Y$. The representation of f is given by 4.4.10.

2. THEOREM. *With the hypotheses of Theorem 1,* Y *is conull if and only if* $z \in W_b(X)$.

Sufficiency: Let $f \in Y'$. By Theorem 1, $f(1) = \Sigma\, \alpha_k + g(z)$. Now by definition of W, $g(z) = \Sigma\, z_k g(\delta^k) = \Sigma\, g(z\cdot\delta^k)$. Meanwhile $f(\delta^k) = \alpha_k + g(z\cdot\delta^k)$. Hence $f(1) = \Sigma\, f(\delta^k)$ and Y is conull.

Necessity. Let $g \in X'$ and define f by $f(y) = g(z\cdot y)$. Then $g(z) = f(1) = \Sigma\, f(\delta^k) = \Sigma\, g(z\cdot\delta^k) = \Sigma\, g(z_k\delta^k) = \Sigma\, z_k g(\delta^k)$ and so $z \in W$. That $z \in \ell^\infty$ is part of the hypothesis.

Lemma 5.6.8 is a special case of Theorem 2.

3. LEMMA. *Let* X *be a conservative* FK *space and* $z \in W_b$. *Then there exists* r *such that* $X \supset z\cdot\Omega(r)$.

Let $Y = z^{-1}\cdot X$. By Theorem 2, Y is conull and the result is immediate from 6.4.2.

4. COROLLARY. *If* X *is a conservative* FK *space with* $W_b \not\subset c_o$, $X \cap \ell^\infty$ *is a non-separable subspace of* ℓ^∞.

From Theorem 3 and 6.3.9.

Now we can obtain a strong improvement of Theorem 6.1.1:

5. THEOREM. *Let* X *be a conservative* FK *space in which* c *is not closed. Then there exists* $z \in \ell^\infty \backslash c_o$ *and a sequence* r *such that* $X \supset z\cdot\Omega(r)$. *In particular* $X \cap \ell^\infty$ *is a non-separable subspace of* ℓ^∞.

It is sufficient by Lemma 3 and Corollary 4 to show that W contains a bounded divergent sequence, or, what is the same, that $2_\infty c_o$ contains one (6.2.6). Referring to the proof of 6.1.1, let $y^m = \sum_{k=1}^{m} x^n$ so that $y^m \in c_o$, $y^m \to x$ and $\|y^m\|_\infty \leq 1 + \Sigma\, 2^{-n-2}$ by the same proof which showed that x is bounded in 3.4.4.

The result of Theorem 5 is new only for coregular spaces; for conull spaces it is contained in 6.4.2. ($z = 1$.)

6. THEOREM. *Let* A *be a conservative matrix which sums a bounded divergent sequence. Then there exists* $z \in \ell^\infty \backslash c_o$ *and a sequence* r *such that* $c_A \supset z\cdot\Omega(r)$. *In particular* $c_A \cap \ell^\infty$ *is a non-separable subspace of* ℓ^∞.

By 6.1.3 c is not closed in c_A and the result follows by Theorem 5.

The FK program fails for Theorem 6 as pointed out in 6.1.4.

6.6. SUBSPACES OF ℓ^∞.

Although ℓ^∞ is too big to be included in c_A for any regular or coregular matrix by 3.5.5, and the same is true for certain subspaces of ℓ^∞ by 6.1.2, there are other subspaces of ℓ^∞ which can be included in such c_A:

1. THEOREM. *Let E be a separable subspace of ℓ^∞. Then there exists a regular triangle A such that $c_A \supset E$. Also, given any matrix $B \in \Phi$, there is a row submatrix A of B such that $c_A \supset E$.*

A *row submatrix* of B is a matrix obtained by deleting rows of B. If B is chosen regular, then A will be regular also. The first part of the theorem follows from the second by 3.5.1. Now let $B \in \Phi$; we may assume $\|B\| = 1$. Define $u_n \in E'$ by $u_n(x) = (Bx)_n$; $|u_n(x)| \leq |(Bx)_n| \leq \|x\|_\infty$. Thus each $u_n \in D$, the unit disc in E'. It follows from 6.0.2 that $\{u_n\}$ has a subsequence $\{u_{k(n)}\}$ with the property that $\lim u_{k(n)}(x)$ exists for each $x \in E$. This says $E \subset c_A$ where A is the row submatrix of B gotten by deleting all but the rows numbered $k(n)$.

2. COROLLARY. *Every matrix in Φ has a row submatrix in Γ.*

Apply Theorem 1 to $E = c$.

The same reasoning can be applied to reflexive BK spaces of bounded sequences, such as ℓ^2. However, any reflexive BK space is separable (6.0.1) so this is nothing new.

6.7. BIGNESS

We have seen only one "absolute" measure of bigness, namely not having a growth sequence (4.2.9). An FK space is included in a BK space if and only if it has a growth sequence (4.2.11). In another context we have seen that conull spaces are "relatively" bigger than coregular ones; every space is included in ω, which is conull, and no conull space can be included in a coregular one, (4.6.3). However, there is nothing absolute here. For example if A is a conull triangle, the map $A: c_A \to c$ is a Banach space equivalence between a conull and a coregular space. If $z = \{(-1)^n\}$ then $z \cdot X$ and X are surely the "same" size, yet one can be coregular and one conull, for example let $(Ax)_n = (-1)^n(x_{n-1}-x_n)$, $(Bx)_n = x_{n-1} + x_n$; then $c_B = z \cdot c_A$. As a third example, two coregular spaces X and Y may be so large that $X + Y = \omega$, for example let $(Ax)_n = x_{2n}$, $(Bx)_n = x_{2n+1}$, $X = c_A$, $Y = c_B$. Paradoxically $X \cap Y$ must be large i.e. there is a lot of "waste" in $X + Y$, for example if $X \cap Y = \{0\}$, X would be closed in $X + Y$, (4.5.1), so $X + Y$ would be coregular (4.6.3). (In fact $X \cap Y \supset c$.)

Finally there are conull spaces which have a growth sequence, e.g. c_A with A a triangle; and coregular spaces with no growth sequence e.g. the space X given in the preceding sentence.

The example just given, of coregular X with $Y = z \cdot X$ conull, shows that coregular and conull are "shape" rather than "size" properties; more specifically, they measure the relationship between 1 and ϕ; in X, 1 is not weakly $\Sigma\, \delta^n$ but z is weakly $\Sigma\, z_n \delta^n$; 1 does not have weak AK but z does; $1 \notin W$, $z \in W$.

CHAPTER 7
SEQUENCE SPACES

7.0. FUNCTIONAL ANALYSIS.

1. Let $q: X \to Y$ be linear and onto where X is a Banach space and such that $q^{\perp}$ is closed in X. Then Y can be made into a Banach space such that q is continuous and open, and a linear functional f on Y is continuous iff $f \circ q$ is continuous on X. [80], Corollary 6-2-15, Theorems 6-2-8, 6-2-5.

2. A pointwise bounded sequence (1.0.3) of continuous linear maps from a Fréchet space to any locally convex space must be equicontinuous. [80], Theorem 9-3-4, Example 9-3-2.

3. The convergence lemma (1.0.5) holds assuming that $\{f_n\}$ is equicontinuous on a topological vector space. [80], #9-3-104.

4. (Open Mapping Theorem). A continuous linear map from a Fréchet space onto a Fréchet space must be an open map. [80], Theorem 5-2-4.

5. Let X, Y be Banach spaces and $T : X \to Y$ a continuous linear bijection. Then for each $f \in X'$ there exists $g \in Y'$ such that $f(x) = g(Tx)$ for $x \in X$. [Proof: Defining g by this formula, let $y = Tx$. Then $|g(y)| = |f(x)| \leq \|f\| \cdot \|x\| \leq \|f\| \cdot \|T^{-1}\| \cdot \|y\|$. Now $\|T^{-1}\| < \infty$ by the open mapping theorem so $g \in Y'$ (1.0.1).]

7.1. MONOTONE NORMS.

In this chapter we present only those results from the extensive theory of sequence spaces which will be needed here. To follow the subject further one may consult [31] and [55], where many of the results are given in greater generality.

1. DEFINITION. *Let* X *be a* BK *space. Then* X *is said to have monotone norm if* $\|x^{(m)}\| \geq \|x^{(n)}\|$ *for* $m > n$ *and* $\|x\| = \sup \|x^{(m)}\|$.

2. EXAMPLE. Let $X = \ell \oplus 1$. (4.5.5.). Then $\|x^{(m)}\| = \sum_{k=1}^{m} |x_i|$, $\|1^{(n)}\| = n$ so this norm satisfies the first condition but not the second. It is not a monotone norm.

3. EXAMPLE. The spaces $c_0, c, \ell^\infty, cs, bs$ have monotone norms.

The next result shows that an AK space may be assumed to have a monotone norm.

4. THEOREM. *Let* X *be a* BK *space with* AK, $\|x\|^* = \sup \|x^{(n)}\|$ *for* $x \in X$. *Then* $\|\cdot\|^*$ *is a monotone norm for* X *and is equivalent with* $\|\cdot\|$. *In particular* $\|x^{(n)}\| \leq K\|x\|$ *for all* $x \in X$.

For example, $\|x+y\|^* = \sup \|x^{(n)}+y^{(n)}\| \leq \sup(\|x^{(n)}\|+\|y^{(n)}\|) \leq \|x\|^* + \|y\|^*$. Now $\|x^{(n)}\| \to \|x\|$ by AK so $\|x\|^* \geq \|x\|$. Conversely let $f_n: X \to X$ be defined by $f_n(x) = x^{(n)}$. Each f_n is continuous since X is an FK space; and the sequence $\{f_n\}$ is pointwise convergent. By uniform boundedness (1.0.3), $\|f_n\| < M$. Hence $\|x^{(n)}\| = \|f_n(x)\| \leq M\|x\|$ and so $\|x\|^* \leq M\cdot\|x\|$. This proves that the norms are equivalent. To prove monotonicity: if $m > n$, $\|x^{(m)}\|^* \geq \|x^{(n)}\|^*$ since each is the sup of an eventually constant sequence agreeing in the first n terms. Also $\sup \|x^{(m)}\|^* = \sup_m \sup\{\|x^{(n)}\| : n \leq m\} = \|x\|^*$.

5. Theorem 4 could be proved without using continuity of coordinates by showing that $\|\cdot\|^*$ is complete and applying the closed graph theorem. Since coordinates are obviously continuous in this norm the same is true in the original norm. This is an important result in functional analysis which has no significance here. See [2] pp. 110-111, [79], Theorem 11.4.1.

6. EXAMPLE. *Let* $A = (C,1)$. *Then* c_A *has monotone norm.* Since A is a triangle, this is a BK space (3.2.2). Fix x and write $u(m,n) = |n^{-1} \sum_{k=1}^{m} x_k|$. Then $\|x\| = \sup u(n,n)$. Also $|(Ax^{(m)})_n| = u(n,n)$ if $n \leq m$, $u(m,n)$ if $n \geq m$. Since $u(m,n)$ is decreasing in n, it follows that $\|x^{(m)}\| = \sup\{u(n,n) : n \leq m\}$.

7.2. DUALS

In this section we discuss various duals and give some inclusion relations.

1. Recall the definitions of X^β, X^γ given in 4.3.5, 4.3.17. These definitions make sense if X is any set of sequences, for example $X^\gamma = \bigcap\{x^\gamma : x \in X\}$ where $x^\gamma = \{y : \Sigma\, x_i y_i$ is a bounded series$\}$.

2. THEOREM. *Let* E, E_1 *be sets of sequences. Then* *(i)* $E \subset E^{\beta\beta}$, *(ii)* $E^{\beta\beta\beta} = E^\beta$, *(iii) If* $E_1 \supset E$ *then* $E_1^\beta \subset E^\beta$. *The same results hold for the* γ *dual.*

Parts (i) and (iii) are trivial. By (i), $E^\beta \subset (E^\beta)^{\beta\beta}$; conversely, by (i) and (iii), $E^\beta \supset (E^{\beta\beta})^\beta$. The proofs for the γ-dual are the same.

3. DEFINITION. *Let* X *be an* FK *space* $\supset \phi$. *Then* $X^f = \{\{f(\delta^n)\} : f \in X'\}$.

It is important to observe that whenever X^f is written it is assumed that $X \supset \phi$.

4. THEOREM. *Let X be an FK space $\supset \phi$. Then $X^f = (\overline{\phi})^f$.*

Let $u \in X^f$, say $u_n = f(\delta^n)$. Let g be the restriction of f to $\overline{\phi}$. Then $u_n = g(\delta^n)$. Conversely let $u \in (\overline{\phi})^f$, say $u_n = g(\delta^n)$. By the Hahn-Banach Theorem (3.0.1) extend g to $f \in X'$. Then $u_n = f(\delta^n)$.

5. EXAMPLE. Let $X = c_o \oplus z$ with z unbounded (4.5.5). Then X is a BK space, $X^f = \ell$ by Theorem 4 and 1.0.2; $X^{ff} = \ell^\infty$ (1.0.2) so $X \not\subset X^{ff}$.

However we get the analogue of part of Theorem 2:

6. THEOREM. *If $X \subset Y$ then $X^f \supset Y^f$. If X is closed in Y, then $X^f = Y^f$.*

If $f \in Y'$ then $f|X \in X'$ by 4.2.4. The second part is by Theorem 4.

7. THEOREM. *Let X be an FK space $\supset \phi$. Then (i) $X^\beta \subset X^\gamma \subset X^f$. (ii) If X has AK, $X^\beta = X^f$, (iii) If X has AD, $X^\beta = X^\gamma$.*

Let $u \in X^\beta$ and define $f(x) = ux$ (1.2.1) for $x \in X$. Then $f \in X'$ by the Banach-Steinhaus theorem (1.0.4). Also $f(\delta^n) = u_n$ so $u \in X^f$. Thus $X^\beta \subset X^f$

(ii): Let $u \in X^f$, say $u_n = f(\delta^n)$. For $x \in X$, $f(x) = f(\Sigma x_n\delta^n) = \Sigma x_n u_n$ so $u \in X^\beta$. The opposite inclusion was just proved.

(iii): Let $u \in X^\gamma$ and define $f_n(x) = \sum_{k=1}^{n} u_k x_k$ for $x \in X$. Then $\{f_n\}$ is pointwise bounded, hence equicontinuous (7.0.2). Since $\lim f_n(x)$ exists for all $x \in \phi$, it must exist for all $x \in X$, (7.0.3) i.e. $u \in X^\beta$. The opposite inclusion is trivial.

(i): $X^{\gamma} \subset (\overline{\phi})^{\gamma} = (\overline{\phi})^{\beta}$ (by (iii)) $\subset (\overline{\phi})^{f} = X^{f}$ by Theorem 4.

8. It is easy to give examples in which $X^{\gamma} \neq X^{f}$; one such is Example 5. To give one in which X has AD is harder; it is given in 5.2.14. That $X^{\beta} \neq X^{\gamma}$ is possible is shown in 7.3.5.

The relationship of the duals with X' is less clear since X' is not a sequence space. For example, no sequence space can have the cardinality of $(\ell^{\infty})'$. The plain fact is that the members of X' are functions, not sequences. More on this in Example 11.

9. THEOREM. *Let X be an FK space* $\supset \phi$. *Then* $X^{\beta} \subset X'$ *in the sense that each* $u \in X^{\beta}$ *can be used to represent a function* $\hat{u} \in X'$ *and the map* $u \to \hat{u}$ *is an isomorphism into. If X has AK the map is onto.*

Namely, $\hat{u}(x) = ux$. Then $\hat{u} \in X'$ by the Banach-Steinhaus Theorem (1.0.4). If $\hat{u} = 0$, then $u_n = \hat{u}(\delta^n) = 0$ so $u = 0$ i.e. the map is one to one. If X has AK and $f \in X'$ let $u_n = f(\delta^n)$. Then for all x, $f(x) = f(\Sigma\, x_k \delta^k) = ux$. Thus $u \in X^{\beta}$ and $f = \hat{u}$.

The converse of the last statement is given in 10.5.1.

10. THEOREM. *Let X be an FK space* $\supset \phi$. *Then* X^{f} *is a quotient of* X' *in the sense that the map* $q: X' \to X^{f}$ *given by* $q(f) = \{f(\delta^n)\}$ *is onto. Moreover the diagram* $X^{\beta} \to X'$, $X' \downarrow X^{f}$, $X^{\beta} \searrow X^{f}$ *commutes.*

The first part is trivial. The second means that for $u \in X^{\beta}$, $q(\hat{u}) = u$, using the notation of Theorem 9.

11. EXAMPLE. Let $X = c$. Then $X^{\beta} = X^{f} = \ell$. The embedding of Theorem 9 is not onto, even though there is a sense in which $c' = \ell$ (1.0.2); for $\lim \in X'$, but if $\lim u = ux$ it would follow that $u_n = \lim \delta^n = 0$ so $\lim x \equiv 0$, contradicting

lim 1 = 1. Also the quotient map in Theorem 10 is not one to one since q(lim) = 0.

12. THEOREM. *Let* X *be an* FK *space* $\supset \phi$. *Then* $X^f = X'$ *(the map of Theorem 10 is one to one) iff* X *has* AD.

Sufficiency: If $q(f) = 0$ then $f = 0$ on ϕ hence $f = 0$. Nexessity: If X does not have AD, the Hahn-Banach Theorem (3.0.1) supplies a function $f \in X'$, $f \neq 0$, $f = 0$ on ϕ. Then $q(f) = 0$.

13. Aficionados of duality theory can do a little Hellinger-Toeplitz number on this theorem. Let $Z = \overline{\phi} \subset X$ and let $i: Z \to X$ be inclusion. Then $i': X' \to Z'$ is the quotient map q of Theorem 10. (Note that $Z' = Z^f = X^f$ by Theorems 12 and 4). A standard result (e.g. [80], Lemma 11-1-7) says $Ri = Nq^{\perp}$. (In particular q is one to one iff i is onto — this is Theorem 12 again!) Thus $Z = Ri = Nq^{\perp} = \phi^{\perp\perp} = \overline{\phi}$. This shows that the value of Z is forced.

14. THEOREM. *Let* X *be a* BK *space* $\supset \phi$. *Then* X^f *is a* BK *space*.

The quotient topology by the map of Theorem 10 (See 7.0.1) makes it a Banach space since X' is one and $q^{\perp} = \{f: f = 0 \text{ on } \phi\}$ is closed in X'. Continuity of coordinates follows from 7.0.1 since $P_n \circ q: X' \to K$ satisfies $|P_n[q(f)]| = |f(\delta^n)| \leq \|f\| \cdot \|\delta^n\|$ hence is continuous.

We saw in Example 5 that X need not be included in X^{ff}. However part of it is:

15. THEOREM. *Let* X *be a* BK *space* $\supset \phi$. *Then* $X^{ff} \supset \overline{\phi}$ *(closure in* X*). Hence if* X *has* AD, $X \subset X^{ff}$.

The second part is equivalent to the first by Theorem 4 so we shall assume that X has AD. But first it is necessary to show that $X^f \supset \phi$ in order for X^{ff} to be meaningful. This is true because $P_k \in X'$ where $P_k(x) = x_k$ since X is an FK space, and $q(P_k) = \delta^k$ (Theorem 10.) Now let $x \in X$. Let $f(u) = u(x)$ for $u \in X'$. Then $|f(u)| \leq \|u\| \cdot \|x\|$ so $f \in X''$ (1.0.1). Apply 7.0.5 to $q: X' \to X^f$ (7.2.12): this yields $g \in (X^f)'$ with $u(x) = g(qu)$ for all $u \in X'$. In particular $x_k = P_k(x) = g(qP_k) = g(\delta^k)$ so $x \in X^{ff}$.

16. EXAMPLE. Let $X = c_0 \oplus z$ (4.5.5) with $z \in \ell^\infty$. Then *$X^{ff} = \ell^f = \ell^\infty \supset X$ but X does not have AD.*

7.3. NINE SPACES

1. EXAMPLE. The familiar BK spaces, bs, cs, c_0, c, ℓ^∞ and ℓ have been introduced. It was pointed out in 1.3.2 that $(\ell^\infty)^\beta = \ell$. (Consider one row of A). That $c_0^\beta = \ell$ is proved similarly. Hence $c^\beta = \ell$. Also $\ell^\beta = \ell^\infty$ [$\supset$: Trivial; $\subset$: Let $y \in \ell^\beta$, $f(x) = xy$. Then $|y_n| = |f(\delta^n)| \leq \|f\| \cdot \|\delta^n\| = \|f\|$. Hence $y \in \ell^\infty$.] Finally $c_0^f = \ell$ and $\ell^f = \ell^\gamma = \ell^\beta = \ell^\infty$ by 7.2.7.

To introduce bv and bv_0 we prove:

2. LEMMA. *If $\Sigma |x_k - x_{k+1}| < \infty$ then $x \in c$.*

For $m \geq n$, $|x_{m+1} - x_n| = |\sum_{k=n}^{m} (x_k - x_{k+1})| \leq \sum_{k=n}^{\infty} |x_k - x_{k+1}| \to 0$.

3. DEFINITION. *bv, the space of sequences of bounded variation, is the set of all x such that the series in Lemma 2 converges; $\|x\|_{bv} = |\lim x| + \Sigma |x_k - x_{k+1}|$; $bv_0 = bv \cap c_0 = \{x \in bv: \lim x = 0\}$.*

4. bv_0 is equivalent with ℓ under the map $x \to y = \{x_n - x_{n+1}\}$, the inverse map being $y \to x$ with $x_n = \sum_{k=n}^{\infty} y_k$. Moreover

$\|x\|_{bv} = \|y\|_1$ for $x \in bv_o$. Thus *bv_o is a Banach space. That it is a BK space* is proved as follows: $x_n = \sum_{k=n}^{m} (x_k - x_{k+1}) + x_{m+1} \to \sum_{k=n}^{\infty} (x_k - x_{k+1})$. Thus $|x_n| \le \|x\|$ for each n.

Hence $bv = bv_o \oplus 1$ is also a BK space (4.5.5). Note that $bv_o = \lim^{\perp}$ has codimension 1 in bv (5.0.1).

5. THEOREM. *(i) bv_o is an AK space. (ii) $bv_o^{\beta} = bv_o^{\gamma} = bv_o^{f} = bs$, (iii) $bv^{\beta} = cs$, (iv) $bv^{\gamma} = bv^{f} = bs$, (v) $cs^{\beta} = cs^{\gamma} = cs^{f} = bv$, (vi) $bs^{\beta} = bv_o$, (vii) $bs^{\gamma} = bs^{f} = bv$.*

(i): $\|x - x^{(m)}\| = \sum_{k=m+1}^{\infty} |x_k - x_{k+1}| \to 0$.

(ii): $bv_o^{\gamma} = bv_o^{\beta} = bv_o^{f}$ by 7.2.7. Also $bv_o^{\beta} \supset bs$ by Dirichlet's test for convergence. Conversely let $u \in bv_o^{f}$, say $u_k = f(\delta^k)$. Then $|\sum_{k=1}^{n} u_k| = |f(1^{(n)})| \le \|f\| \cdot \|1^{(n)}\| = \|f\|$, so $u \in bs$.

(iii): $1 \in bv$ hence $bv^{\beta} \subset 1^{\beta} = cs \subset bv^{\beta}$. The last inclusion is by Abel's test for convergence.

(iv): $bv^{\gamma} = (bv_o \oplus 1)^{\gamma} = bv_o^{\gamma} \cap 1^{\gamma} = bs$ by (ii). The rest is by 7.2.6 and (ii).

(v): $cs^{f} = cs^{\gamma} = cs^{\beta} = bv^{\beta\beta}$ by (iii) and 7.2.7. Conversely, let $u \in cs^{\gamma}$. Now if $y \in c$ and $x_n = y_n - y_{n-1}$ (convention: $y_o = 0$) then $x \in cs$ and so $u \cdot x \in bs$. By Abel's identity (1.2.9), $\sum_{k=1}^{n} u_k x_k = \sum_{k=1}^{n-1} (u_k - u_{k+1}) y_k + u_n y_n = (Ay)_n$ where $a_{nk} = u_k - u_{k+1}$ for $k < n$, u_n for $k = n$, 0 for $k > n$. Since $u \cdot x \in bs$ it follows that $Ay \in \ell^{\infty}$ and since this is true for each $y \in c$, $\|A\| < \infty$ (1.3.3). Thus for each n, $\|A\| \ge \sum_{k=1}^{n-1} |u_k - u_{k+1}|$ and so $u \in bv$.

(vi) $bs^{\beta} = bv_o^{\beta\beta} \supset bv_o$ by (ii). Conversely, $bs^{\beta} \subset cs^{\beta} = bv$ by (v); finally let $u = \{(-1)^n\}$. Then $u \in bs$ so $bs^{\beta} \subset u^{\beta} \subset c_o$.

(vii) $bs^{\gamma} = bv^{\gamma\gamma} \supset bv$ by (iv). Conversely, $bs^{\gamma} \subset cs^{\gamma} = bv$ by (v). The other part is by 7.2.4 and (v).

The ℓ^p spaces for $p > 1$ play an insignificant role in this book. We give a brief discussion with references.

6. DEFINITION. $\|x\|_p = (\Sigma|x_n|^p)^{1/p}$ *for* $p \geq 1$; $\ell^p = \{x : \|x\|_p < \infty\}$.

7. EXAMPLE. ℓ^p *is a BK space with AK for* $p \geq 1$. *For* $p > 1$, *let* $1/p + 1/q = 1$ *then* $f \in (\ell^p)'$ *if and only if there exists* $y \in \ell^q$ *such that* $f(x) = xy$; *moreover* $\|f\| = \|y\|_q$; *also* $(\ell^p)^{\beta} = \ell^q$.

See [79], Examples 4.1.3, 6.4.3. Completeness follows from reflexivity.

8. The inequality $|f(x)| \leq \|f\|\cdot\|x\|$ (1.0.1) yields, by Example 7, *Hölder's inequality:* $|\Sigma\, x_i y_i| \leq \|x\|_p \|y\|_q$, the left side being absolutely convergent if the right side is finite.

9. THEOREM. *Each of the nine spaces has monotone norm.*

Let $x \in bv$. Then

$$\|x^{(n)}\| = \sum_{k=1}^{n-1} |x_k - x_{k+1}| + |x_n| \tag{1}$$

Also for $m > n$, $|x_n| = |\sum_{k=n}^{m-1} (x_k - x_{k+1}) + x_m| \leq \Sigma\, |x_k - x_{k+1}| + |x_m|$. Substituting this in (1) and using (1) with n replaced by m gives $\|x^{(n)}\| \leq \|x^{(m)}\|$. Also from (1), $\|x\| = \lim \|x^{(n)}\|$. Thus bv has monotone norm. This implies the same for bv_o. For the other seven spaces the result is obvious.

7.4. DETERMINING SETS.

1. DEFINITION. *Let* X *be a BK space. Then* $D = D(X) = \{x \in \phi : \|x\| \leq 1\}$. (We do not assume $X \supset \phi$).

This is not the unit sphere in X, but rather its intersection with ϕ.

2. DEFINITION. *Let* X *be a* BK *space. A subset* E *of* ϕ *will be called a determining set for* X *if* $D(X)$ *is the absolutely convex hull of* E.

3. EXAMPLE. $\{\delta^n\}$ *is a determining set for* ℓ. Let A be the absolutely convex hull of this set. If $x \in A$, $x = \sum_{k=1}^{m} t_k \delta^k$ so $x \in \phi$ and $\|x\| \leq \Sigma|t_k|\cdot\|\delta^k\| = \Sigma|t_k| \leq 1$. Conversely if $x \in D$, $x = \sum_{k=1}^{m} x_k\delta^k$ and $1 \geq \|x\| = \Sigma|x_k|$; thus $x \in A$.

4. EXAMPLE. $\{1^{(n)}\}$ *is a determining set for* bv_o (see 4.2.13 for the notation). Let A be the absolutely convex hull of this set. If $x \in A$, $x = \sum_{k=1}^{m} t_k 1^{(k)}$ so $x \in \phi$ and $\|x\| \leq \Sigma|t_k|\cdot\|1^{(k)}\| = \Sigma|t_k| \leq 1$. Conversely if $x \in D$, $x = \sum_{k=1}^{m} x_k\delta^k = \Sigma\, x_k(1^{(k)}-1^{(k-1)})$ (Convention: $1^{(0)} = 0$). By Abel's identity (1.2.9) $x = \sum_{k=1}^{m} (x_k-x_{k+1})1^{(k)}$ since $x_{m+1} = 0$. Also $\Sigma|x_k-x_{k+1}| \leq \|x\| \leq 1$ so $x \in A$.

5. EXAMPLE. *Let* E *be the set of all sequences in* ϕ *each of whose non-zero terms is* ± 1. *Then* E *is a determining set for* c_o. Let A be the absolutely convex hull of this set. If $x \in A$, $x = \sum_{k=1}^{m} t_k s^k$ so $x \in \phi$ and $\|x\| \leq \Sigma\, |t_k|\cdot\|s^k\| = \Sigma\, |t_k| \leq 1$. Conversely if $x \in D$, $x = \sum_{k=1}^{m} x_k\delta^k$; suppose first that $|x_i| \geq |x_{i+1}|$ for all i. Let $\varepsilon_i = \operatorname{sgn} x_i$, $s^j = \sum_{i=1}^{j} \varepsilon_i\delta^i$ for $j = 1,2,\dots,m$. Each $s^j \in E$ and $x = \sum_{j=1}^{m} (|x_j|-|x_{j+1}|)s^j \in A$. In general, let y be x rearranged so that $|y_i| \geq |y_{i+1}|$ and express y as a

member of A. Since E is invariant under permutation of the terms of its members, so is A; hence $x \in A$.

CHAPTER 8
INCLUSION AND MAPPING

8.0. FUNCTIONAL ANALYSIS

1. A set in a topological vector space is called *bounded* if it is absorbed by every neighborhood of 0. A continuous linear map preserves bounded sets; hence a bounded set remains bounded in any smaller topology. If A, B are bounded sets, A + B is bounded. [80], §4-4.

2. In a locally convex space the absolutely convex hull of a bounded set is bounded. Every set which is bounded in the weak topology is bounded i.e. f [E] bounded for all $f \in X'$ implies E is bounded in X. See 4.0.11 and [80] #7-1-1, Theorem 8-4-1.

3. If a linear map from a metrizable topological vector space preserves bounded sets it must be continuous. [80], Theorem 4-4-9, Example 4-4-7.

4. A set E in a locally convex space is bounded iff p [E] is bounded for each continuous seminorm. [80] Theorem 7-2-6.

8.1

TABLE

FROM / TO	bv_0 (8.6.4)	bv	bs	c_0 (8.6.2)	c	cs (8.6.5)	ℓ (8.6.3)	$\ell^p, p>1$ (8.6.6)	ℓ^∞
bv_0	8.4.2C	8.4.2C	8.5.6B	8.4.3A	8.4.7A	8.5.5B	8.4.1C	8.5.10	8.4.7A
bv	8.4.2C	8.4.2C	8.5.6A	8.4.3A	8.4.7A	8.5.5B	8.4.1C	8.5.10	8.4.7A
bs	8.4.2B	8.4.2B	8.4.6C	8.4.6A	8.4.6A	8.4.6B	8.4.1B	8.4.6D	8.4.6A
c_0	8.4.2A	8.4.2A	8.5.6E	8.4.5A	1.3.6 8.4.5A	8.4.5B	8.4.1A	8.4.5D	1.7.19
c	8.4.2A	8.4.2A	8.5.6D	8.4.5A	1.3.6 8.4.5A	8.4.5B	8.4.1A	8.4.5D	1.7.18
cs	8.4.2B	8.4.2B	8.5.9	8.4.6A	8.4.6A	8.4.6B	8.4.1B	8.4.6D	8.5.8
ℓ	8.4.2D	8.4.2D	8.4.9B 8.5.6D	8.4.3B	8.4.9A	8.4.7B 8.5.5A	8.4.1D	8.4.8B	8.4.9A
$\ell^p, p>1$	8.4.2D	8.4.2D	8.5.6C	8.4.3B	8.4.8A	8.5.5A	8.4.1D	8.5.12	8.4.8A
ℓ^∞	8.4.2A	8.4.2A	8.4.5C	1.3.3 8.4.5A	8.4.5A	8.4.5B	8.4.1A	8.4.5D	1.3.3 8.4.5A
ω	8.4.2	8.4.2	8.4.5C	8.4.5A	8.4.5A	8.4.5B	8.4.1	8.4.5D	8.4.5A

8.1. INTRODUCTION

In this chapter we give a general principle which will yield, for a given FK space, criteria for an FK space to include it. From these criteria we obtain conditions on a matrix A equivalent to $A \in (X:Y)$ i.e. A maps X into Y, this is done by simply applying the inclusion criteria to decide when $Y_A \supset X$. The classical treatment of these mapping theorems was to guess the conditions e.g. Silverman-Toeplitz (1.3.6), showing them sufficient by some standard inequality, and then (usually the difficult part) showing them necessary by contradiction: if A does not satisfy the conditions, a sequence $x \in X$ is constructed such that $Ax \notin Y$. The construction might be by a gliding hump, possibly extremely complicated. For a very inclusive list of references see [72].

The soft approach, initiated in Banach's book [2] in 1932, is the one we showed in 1.3.3, 1.3.6, it is based on uniform boundedness. The approach of this chapter depends ultimately on the closed graph theorem. Whenever this is true, calculations tend to be easier. The approach was worked out in a series of seminars at Lehigh University involving Wilansky, Snyder, Bennett, Kalton, Garling, Ruckle and others. It overlaps with ideas of many other writers.

The success of the method may be judged in that it gives short proofs *and derivations* (the classical methods gave only proofs) of 81 conditions out of a possible 90 in the Table, p. 116. Of the remainder, 8 are $(\ell^\infty:c_o)$ and its modifications; no soft treatment of these is known. Finally there is the famous pair $(\ell^p:\ell^r)$ about which we say very little. It is interesting that ℓ^2, the best of spaces, is the least tractable in the present context. The reason, ultimately, seems to be that its unit disc has too many extreme points!

8.2. INCLUSION

In order for Y to include X it is necessary that the topology of Y be smaller than that of X (4.2.4). A converse result can be obtained as follows:

1. LEMMA. *Let X be an AD space and Y an FK space $\supset \phi$. Suppose that $T_Y|\phi \subset T_X|\phi$. Then $Y \supset X$.*

Let $x \in X$. There exist $u^n \in \phi$ with $u^n \to x$ in X. Then $\{u^n\}$ is a Cauchy sequence in Y and so $u^n \to y$ in Y. Since X, Y are FK spaces, $u^n \to$ both x and y in ω, hence $x = y$ and so $x \in Y$.

For the concept "bounded" refer to 8.0.1.

2. LEMMA. *Let X be an AD space and Y an FK space $\supset \phi$. Suppose that every subset of ϕ which is bounded in X is also bounded in Y. Then $Y \supset X$.*

The inclusion map: $(\phi, T_X) \to Y$ preserves bounded sets hence is continuous (8.0.3). This is equivalent to the hypothesis of Lemma 1.

If X is not AD these arguments may be applied to the AD space $\overline{\phi}$. Also the properties of ϕ do not enter. Here is a general statement of what was proved. (This all holds for FH spaces too.):

3. LEMMA. *Let X, Y be FK spaces and let $E \subset X \cap Y$. These are equivalent (i) $T_Y|E \subset T_X|E$, (ii) every subset of E which is bounded in X is also bounded in Y, (iii) $Y \supset cl_X E$.*

(i) = (ii): (i) is true iff the inclusion map is continuous; this in turn, is true iff this map preserves bounded sets (8.0.1, 8.0.3).

(iii) implies (i): By the inclusion theorem 4.2.4 applied to FK spaces Y and $\overline{E}$.

(i) implies (iii): The proof of Lemma 1 applies with ϕ replaced by E, X by $\overline{E}$.

4. THEOREM. *Let* X *be a BK space with AD,* Y *an FK space* $\supset \phi$, *and* E *a determining set for* X *(7.4.2). Then* $Y \supset X$ *iff* E *is bounded in* Y.

Necessity is by 4.2.4 and 8.0.1. To prove sufficiency, note first that $D = D(X)$ is bounded, in Y by 8.0.2. Now, applying Lemma 2, let B be a subset of ϕ which is bounded in X. Then $B \subset mD$ for some m, hence B is bounded in Y.

5. EXAMPLE. $Y \supset \ell$ *iff* $\{\delta^n\}$ *is bounded in* Y. By Theorem 4 and 7.4.3. (Another treatment in 8.6.3.)

Theorem 4 uses local convexity of Y. For general Y we have $Y \supset X$ if and only if $D(X)$, the absolutely convex hull of E, is bounded in Y. In fact, Example 5 fails if Y is not assumed locally convex since $\{\delta^n\}$ is bounded in $\ell^{1/2}$. W. H. Ruckle has shown, [56], that $\cap\{\ell^p : p > 0\}$ is the smallest locally semiconvex FK space in which $\{\delta^n\}$ is bounded, and that the intersection of all FK spaces in which $\{\delta^n\}$ is bounded is ϕ, hence there is no smallest. He has also shown, [57], Theorem 4.1 that for each balanced and bounded set B in ω there is a BK space which is the smallest (locally convex) FK space in which B is a bounded set.

6. EXAMPLE. $Y \supset bv_0$ *iff* $\{1^{(n)}\}$ *is bounded in* Y. By Theorem 4 and 7.4.4. (Another treatment in 8.6.4).

7. EXAMPLE. Let E be the set of all finite sequences of 0's and 1's. Thus a member of E is the characteristic function

of a finite set of positive integers. *Then* $Y \supset c_o$ *iff* E *is bounded in* Y. (Another treatment in 8.6.2.) This is by Theorem 4 and 7.4.5 except for one detail: E is not the determining set, call it E_0, given in 7.4.5. However $E \subset E_0 = E - E$ (i.e. each member of E_0 is the difference of two members of E) so E is bounded if and only if E_0 is (8.0.1). The condition that E be bounded is referred to in the literature as: 1 has unconditional section boundedness.

8. Since ℓ^∞ is not an AD space none of the techniques given here applies to it. Conditions for spaces Y_A, with particular Y, to include ℓ^∞ i.e. $A \in (\ell^\infty : Y)$ are referred to in the Table p. 116. A sufficient condition of some interest is given in [80], Example 15.2.2, namely if Y includes all sequences of 0's and 1's then $Y \supset \ell^\infty$. (Compare 19.3.4).

8.3. MAPPING THEOREMS

A mapping theorem is a characterization of (X:Y) i.e. of matrices $A: X \to Y$. An example is the Silverman-Toeplitz characterization of (c:c), 1.3.6.

For a matrix A let $Y_A = \{x \in \omega_A : Ax \in Y\}$; then $A \in (X:Y)$ if and only if $Y_A \supset X$. Hence inclusion theorems lead to mapping theorems.

The presentation of results will be as follows: All 9 spaces in the top row of the Table p. 116 are BK spaces.

(a) Direct application of the principle $A \in (X:Y)$ if and only if $Y_A \supset X$. This is restricted to the cases in which X is an AK space. The end result is Theorem 4.

(b) Varying Y when (X:Y) is known and X has AK, Remark 6.

(c) Enlarging X by one dimension; Remark 7.

(d) Obtaining $(Y^\beta : X^\beta)$ from $(X:Y)$, Theorem 10.

The display of applications in the next section will follow this order

The inclusion Theorem 8.2.4 shows that we need criteria for boundedness of sets in Y_A, an FK space by 4.3.12. (This reference will be omitted from now on.)

1. LEMMA. *Let Y be an FK space, A a matrix. A subset E of Y_A is bounded iff E is bounded in ω_A and $A[E]$ is bounded in Y.*

Necessity: Since $Y_A \subset \omega_A$, E is bounded in ω_A by 4.2.4, 8.0.1. Also $A: Y_A \to Y$ is continuous, hence preserves bounded sets. (8.0.1.)

Sufficiency: It is sufficient to prove that $f[E]$ is bounded for all $f \in Y_A'$. (8.0.2.) Such $f = F + g \circ A$ where $F \in \omega_A'$, $g \in Y'$. (4.4.2.) Trivially $F[E]$ and $g \circ A[E] = g[A[E]]$ are bounded. We could also have used 8.0.4 and 4.3.12 to give a direct proof.

2. THEOREM. *Let X be an FK space with AD, Y an FK space and A a matrix. Then $A \in (X:Y)$ iff (i). The columns of A belong to Y, (ii). $A[E]$ is bounded in Y for each $E \subset \phi$ which is bounded in X, (iii). The rows of A belong to X^β. The condition (ii) may be replaced by (ii)'. $A:(\phi, T_X) \to Y$ is continuous.*

(ii) = (ii)' by 8.0.1, 8.0.3.

Necessity: Condition (iii) is simply the statement $\omega_A \supset X$. Condition (i) says $Y_A \supset \phi$ which is certainly necessary. Finally (ii)' is immediate, indeed $A: X \to Y$ is continuous (4.2.8).

Sufficiency: We have to prove $Y_A \supset X$; we shall use the criterion of 8.2.2. Let E be a subset of ϕ which is bounded in X. It is required to prove that E is bounded in Y_A. Lemma 1 applies: E is bounded in X hence in ω_A (8.0.1) since $X \subset \omega_A$ by (iii), and $A[E]$ is bounded in Y by (ii).

Under a more stringent assumption on X or A we can reduce the number of conditions in Theorem 2. The result is Theorem 4. In Example 5 it is shown that the extra assumptions cannot be dropped. We begin by reducing the conditions of Lemma 1 for a particular choice of E, Y.

3. LEMMA. *Let X be a BK space with AK and A a matrix. Then $D = D(X)$ (7.4.1) is bounded in ω_A iff $A[D]$ is bounded in ω.*

Necessity is trivial since $A: \omega_A \to \omega$ is continuous. To prove sufficiency, note first that $D \subset \phi \subset \omega_A$. The seminorms p_n, h_n of ω_A are given in 4.3.8. First D is bounded in X so each $p_n[D]$ is bounded since X is an FK space. Next, fixing n, for any m and $x \in D$ we have $|\sum_{k=1}^{m} a_{nk}x_k| = \sum_{k=1}^{\infty} a_{nk}x_k^{(m)}| = p_n(Ax^{(m)})$. There exists K such that $\|x^{(m)}\| \leq K$ (7.1.4), so $Ax^{(m)} \in A[K \cdot D]$, a bounded set in ω by hypothesis. It follows that $p_n(Ax^{(m)}) < C$ where C is a constant depending on n (which is fixed in this argument.) From this, $h_n(x) \leq C$ i.e. $h_n[D]$ is bounded. This completes the proof that D is bounded in ω_A.

4. THEOREM. *Let X be a BK space, and E a determining set for X (7.4.2). Let Y be an FK space and A a matrix. Suppose that either X has AK or A is row-finite. Then $A \in (X:Y)$ iff (i). The columns of A belong to Y, and (ii). $A[E]$ is a bounded subset of Y.*

Necessity is immediate exactly as in Theorem 2. As to sufficiency: condition (i) of Theorem 2 is the same as here; condition (ii) of Theorem 2 holds, for if $E_1 \subset \phi$ is bounded in X, then $E_1 \subset K.D$ for some K, hence $A[E_1] \subset K.A[D]$ is bounded in Y by (ii) since $A[D]$ is the absolutely convex hull of $A[E]$. Condition (iii) of Theorem 2 is trivial if A is row-finite. Finally let X have AK; $A[E]$ is bounded in Y by (ii), hence in ω. By Lemma 3, E is bounded in ω_A. By 8.2.4, $\omega_A \supset X$ which is just condition (iii) of Theorem 2.

5. EXAMPLE. *Theorem 4 fails if X does not have AK or A is not row-finite* even if X is an AD space of the form c_A with A a conservative triangle. Let X be such a space with $X^\beta \neq X^f$ (5.2.14). Let $u \in X^f \backslash X^\beta$. Let $a_{1k} = u_k$, $a_{nk} = 0$ for $n > 1$. Then $\omega_A \not\supset X$ so $A \notin (X:Y)$ no matter what Y is. However, if Y is any FK space $\supset \phi$, A does satisfy (i) (ii) of Theorem 4. The first is trivial. The second will follow when we show that $A: (\phi, T_X) \to Y$ is continuous since E is bounded in the domain space. Let $f \in X'$ with $f(\delta^k) = u_k$. Then, for $x \in \phi$, $Ax = f(x)\delta^1$ so A is continuous.

6. The following technique called *improvement of mapping* will allow knowledge of $(X:Y)$ to yield $(X:Y_1)$ with Y_1 a closed subspace of Y. *Let X be an AD space, Y and Y_1 FK spaces with Y_1 a closed subspace of Y. Then $A \in (X:Y_1)$ iff $A \in (X:Y)$ and the columns of A belong to Y_1.* Necessity is trivial since $\phi \subset X \subset Y_1)_A$. Sufficiency: $A: X \to Y$ is continuous (4.2.8) and $A[\phi] \subset Y_1$. Since ϕ is dense in X, $A[X] \subset Y_1$

7. Suppose that X is an FK space and $X_1 = X \oplus 1$. Then $A \in (X_1:Y)$ iff $A \in (X:Y)$ and $A1 \in Y$.

Next is an important set of results connecting properties of A with those of A^T, the transposed matrix. This is related to but different from the dual (or adjoint) map $T':Y' \to X'$ (14.0.5).

To illustrate Theorem 8, let $X = \ell \oplus 1$ (4.5.5), $Y = \ell$, $(Ax)_n = x_n - x_{n-1}$. Then $A \in (X:Y)$ but $A^T \notin (Y^\beta : X^\beta) = (\ell^\infty : cs)$ since for $x = (-1)^n$, $A^T x = 2x$.

Also it is shown in 14.5.7 that A^T need not be in $(Y^\beta : X^\gamma)$:

8. THEOREM. *Let X be an FK space and Y any set of sequences. If $A \in (X:Y)$ then $A^T \in (Y^\beta : X^f)$. If X, Y are BK spaces and Y^β has AD, then $A^T \in (Y^\beta : \overline{X^\beta})$ where the closure is taken in X^f. (Improvement of mapping).*

All spaces mentioned in the second part are BK spaces by 4.3.16 and 7.2.14. Let $z \in Y^\beta$ and define $f \in X'$ by $f(x) = z(Ax) = \Sigma z_n(Ax)_n$. This converges by hypothesis and f is continuous by the Banach-Steinhaus Theorem (1.0.4). Then $f(\delta^k) = \Sigma z_n a_{nk} = (A^T z)_k$ so $A^T z \in X^f$. To prove the second part note that $\omega_A \supset X$ so the rows of A belong to X^β; these are the columns of A^T and the result follows by Remark 6.

9. THEOREM. *Let X, Z be BK spaces with AK, $Y = Z^\beta$. Then $(X:Y) = (X^{\beta\beta} : Y)$ and $A \in (X:Y)$ iff $A^T \in (Z : X^\beta)$.*

Note first that X^β is a BK space and is equal to X^f by 4.3.16 and 7.2.7. Let $A \in (X:Y)$. By Theorem 8, $A^T \in (Z^{\beta\beta} : X^\beta) \subset (Z:X^\beta)$. Conversely if $A^T \in (Z:X^\beta)$ then by Theorem 8, $A \in (X^{\beta\beta} : Z^\beta) \subset (X:Z^\beta) = (X:Y)$. This proves the second part. To prove the first part: if $A \in (X:Y)$ then $A^T \in (Z:X^\beta)$ as just proved. By Theorem 8, $A \in (X^{\beta\beta} : Y)$.

10. THEOREM. *The result of Theorem 9 holds if X is any one of the nine spaces listed in the top row of the Table, p. 116. (Recall that $Y = Z^\beta$, where Z is any BK space with AK. Of the nine spaces, Y may be bv, bs, $\ell^p (p \geq 1)$, and ℓ^∞ corresponding to $Z = cs$, bv_o, $\ell^q (1/p + 1/q = 1)$, c_o, and ℓ.)*

The proof given covers the cases when X is an AK space. Also by Theorem 9, $(\ell^{\infty}:Y) = (c:Y) = (c_o:Y)$ and $A \in (c_o:Y)$ if and only if $A^T \in (Z:\ell)$ so Theorem 9 holds for $X = \ell^{\infty}, c$.

Next let $A \in (bs:Y)$. By Theorem 8 this implies that $A^T \in (Y^{\beta}:bv)$, hence $A^T \in (Z:bv)$ since $Z \subset Z^{\beta\beta} = Y^{\beta}$. Now $\omega_A \supset bs$ i.e. each row of A lies in $bs^{\beta} = bv_o$, hence each column of A^T lies in bv_o. By Remark 6, $A^T \in (Z:bv_o)$. Conversely this implies that $A \in (bs:Z^{\beta}) = (bs:Y)$ by Theorem 8. Since $bs^{\beta\beta} = bs$ the other part of Theorem 9 is automatically true.

Finally let $A \in (bv:Y)$. The preceding argument applies with bs, bv, bv_o replaced by bv, bs, cs respectively.

11. The reader may have wondered why the last part of Theorem 9 was not written as $A^T \in (Y^{\beta}:X^{\beta})$ since this is equal to $(Z:X^{\beta})$ by the first part. The reason is that in this form Theorem 10 would not be true (8.4.9).

12. It is trivial that: *given* A, X, *there exists* Y *such that* $Y_A \supset X$ *iff* $\omega_A \supset X$ *and iff the rows of* A *belong to* X^{β}. This gives a set of minimal conditions of which the following is typical: *if the rows of* A *are not all in* ℓ, *there exists no* Y *such that* Y_A *is conservative.*

13. Theorems 9, 10 hold also for $Y = \omega$; indeed, *for any* X, $A \in (X:\omega) = (X^{\beta\beta}:\omega)$ *iff* $A^T \in (\phi:X^{\beta})$ *and if and only if the rows of* A *belong to* X^{β}. If r is any row of A, $r \in X^{\beta}$ implies $r^{\beta} \supset X^{\beta\beta} \supset X$. Conversely $r^{\beta} \supset X$ implies $r \in r^{\beta\beta} \subset X^{\beta}$. Thus $(X:\omega) = (X^{\beta\beta}:\omega)$. Also $A^T \in (\phi:X^{\beta})$ if and only if the columns of A^T (= rows of A) belong to X^{β}. Taking $X = \omega$ we have $A \in (\phi:\phi)$ *iff* $A^T \in (\omega:\omega)$ *iff* A^T *is row-finite i.e. iff* A *is column-finite.*

8.4. EXAMPLES.

We now proceed to list spaces (X:Y) following the order of presentation given at the beginning of §8.3

1. EXAMPLE. Let Y be an FK space. Then *$A \in (\ell:Y)$ iff the columns of A form a bounded set in Y.* We apply 8.3.4 with $E = \{\delta^k\}$ using 7.4.3. The result follows since $A\delta^k$ is the kth column of A. From 8.3.13, *each row of A must be in ℓ^∞ and this is the condition for $A \in (\ell:\omega)$.*

1A. EXAMPLE. From Example 1, *$A \in (\ell:\ell^\infty)$ iff $\sup_{n,k} |a_{nk}| < \infty$. For $(\ell:c_o)$, $(\ell:c)$ add to this the requirement that the columns of A are null, convergent, respectively,* as in 8.3.6. We have actually used this result already! The argument of 6.2.6 proves that the matrix given there lies in $(\ell:c)$.

1B. EXAMPLE. From Example 1, *$A \in (\ell:bs)$ iff $\sup_{m,k} |\sum_{n=1}^{m} a_{nk}| < \infty$. For $(\ell:cs)$ add: $\Sigma_n a_{nk}$ is convergent for each k as in 8.3.6.*

1C. EXAMPLE. From Example 1, *$A \in (\ell:bv)$ iff $\sup_k \sum_n |a_{nk}-a_{n-1,k}| < \infty$.* (Convention: $a_{ok} = 0$.) *For $(\ell:bv_o)$ add,* (8.3.6), *each column of A is null.* (This implies that it is in bv_o since it is in bv by the other condition.)

1D. EXAMPLE. From Example 1, *$A \in (\ell:\ell^p)$, $p \geq 1$, iff $\sup_k \sum_n |a_{nk}|^p < \infty$.* Thus, for example $(C,1) \in (\ell:\ell^p)$ iff $p > 1$.

2. EXAMPLE. To consider $A \in (bv_o:Y)$ apply 8.3.4 with $E = \{1^{(m)}\}$ by 7.4.4. Now $A1^{(m)} = \sum A\delta^k = \sum_{k=1}^{m} a^k$ where a^k is the kth colum of A, so *$A \in (bv_o:Y)$ iff Σa^k is a bounded series in Y,*

$A \in (bv:Y)$ iff $A \in (bv_o:Y)$ and $A1 \in Y$ (8.3.7),

$A \in (bv_0:\omega)$, $(bv:\omega)$ iff the rows of A lie in bs, cs respectively (8.3.13).

2A. EXAMPLE. (Important). From Example 2, *$A \in (bv_0:\ell^\infty)$ iff* $\sup_m \|\sum_{k=1}^m a^k\|_\infty < \infty$ i.e. $\sup_{m,n} |\sum_{k=1}^m a_{nk}| < \infty$. Interchanging m, n we see that *$A \in (bv_0:\ell^\infty)$ iff the set of rows of A is a bounded set in bs.* Another derivation of this is to observe the equivalence of $A^T \in (\ell:bs)$ by 8.3.10 and apply Example 1.

$A \in (bv_0:c_0)$, $(bv_0:c)$ iff $A \in (bv_0:\ell^\infty)$ and the columns of A are null, convergent respectively.

$A \in (bv:\ell^\infty)$ iff the set of rows of A is a bounded set in cs. The easiest way to see this is to apply 8.3.10 and then apply Example 1 to $A^T \in (\ell:cs)$. Finally, (8.3.7), *$A \in (bv:c_0)$ iff $A \in (bv_0:c_0)$ and $A1 \in c_0$, $A \in (bv:c)$ iff $A \in (bv_0:c)$ and $A1 \in c$.*

2B. EXAMPLE. From Example 2, *$A \in (bv_0:bs)$ iff* $\sup_{m,n}|\sum_{k=1}^n \sum_{i=1}^m a_{ik}| < \infty$. *For $(bv_0:cs)$ add: $\sum_n a_{nk}$ is convergent for each k* (8.3.6); *for $(bv:bs)$, $(bv:cs)$ add* (8.3.7) *$A1 \in bs$* (or A has rows in cs), *$A1 \in cs$.*

2C. EXAMPLE. From Example 2, *$A \in (bv_0:bv)$ iff* $\sup_m \sum_n |\sum_{k=1}^m a_{nk}-a_{n-1,k}| < \infty$. *For $(bv_0:bv_0)$ add* (8.3.6) *that each column of A is null* (i.e. in bv_0 as in Example 1C); *for $(bv:bv)$, $(bv:bv_0)$ add* (8.3.7) *$A1 \in bv$* (or A has convergent rows), *$A1 \in bv_0$.*

2D. EXAMPLE. From Example 2, *$A \in (bv_0:\ell^p)$, $p \geq 1$ iff* $\sup_m \sum_n |\sum_{k=1}^n a_{nk}|^p < \infty$. *For $(bv:\ell^p)$ add (8.3.7) $A1 \in \ell^p$* (or A has convergent rows.)

3. EXAMPLE. Let Y be an FK space. Then *$A \in (c_o:Y)$ if and only if $\{A(K): K$ is a finite set of positive integers$\}$ is bounded in Y.* By $A(K)$ we mean the sequence y where $y_n = \sum\{a_{nk}: k \in K\}$; thus $y = Ax$ where x is the characteristic function of K. We apply 8.3.4 with E as in 8.2.7. Also *$A \in (c:Y)$ iff $A \in (c_o:Y)$ and $A1 \in Y$* by 8.3.7. As in Example 2 the last condition can often be replaced by: the rows of A are convergent.

3A. EXAMPLE. From Example 3, *$A \in (c_o:bv)$ iff*

$$\sup\{\sum_n |\sum_{k\in K} a_{nk}-a_{n-1,k}| : K \text{ a finite set of positive integers}\} < \infty.$$

To get $(c_o:bv_o)$ add (8.3.6) *that each column is null* (i.e. in bv_o as in Example 1C). For $(c:bv_o)$ the conditions are $A \in (c_o:bv_o)$ and (8.3.7) $A1 \in bv_o$.

3B. EXAMPLE. From Example 3, *$A \in (c_o:\ell^p)$, $p \geq 1$, iff*

$$\sup\{\sum_n |\sum_{k\in K} a_{nk}|^p : K \text{ a finite set of positive integers}\} < \infty.$$

4. EXAMPLE. We can take advantage of the preceding work to calculate a few β-duals. Let A be a diagonal matrix and set $y = \{a_{nn}\}$. Then $cs_A = y^\beta$ and so, for any space X, $cs_A \supset X$ iff $y^\beta \supset X$, i.e. $y \in X^\beta$. Thus *$A \in (X:cs)$ if and only if $y \in X^\beta$.* Taking $X = \ell$, we have $y \in \ell^\beta$ if and only if $A \in (\ell:cs)$. From Example 1B this gives immediately $\ell^\beta = \ell^\infty$. Taking $X = bv_o$ and using Example 2B gives $bv_o^\beta = bs$. An extension of Example 3 could be used to show $c_o^\beta = \ell$ — we chose to postpone $(c_o:cs)$ to Example 6A for convenience. Use of this would be circular.

We now come to the fourth stage in the order of presentation given at the beginning of §8.3: this is the application of 8.3.10. Several pairs of spaces will fall into more than one category; in such cases we choose the simpler treatment if possible — or choose one at random. For example $(c_o:\ell^\infty)$ could be treated by Example 3, but far more easily by Example 5A.

5. EXAMPLE. Take $Y = \ell^\infty$ in 8.3.10. For X any one of our nine spaces, *$(X:\ell^\infty) = (X^{\beta\beta}:\ell^\infty)$ and $A \in (X:\ell^\infty)$ iff $A^T \in (\ell:X^\beta)$.*

5A. EXAMPLE. From Example 5, $(c_o:\ell^\infty) = (c:\ell^\infty) = (\ell^\infty:\ell^\infty) = \Phi$, (1.3.4). The last equality is by Example 1D applied to A^T. *For $(c_o:c)$, $(c_o:c_o)$ add* (8.3.6) *the condition: the columns of A are convergent, null. To get $(c:c)$ and $(c:c_o)$ add* (8.3.7) *$A1 \in c$, c_o.* This is a complete proof of the Silverman-Toeplitz Theorem 1.3.6.

Note that $(\ell^\infty:c_o)$ and $(\ell^\infty:c)$ are not covered by these techniques.

From 8.3.13, *$(c_o:\omega) = (c:\omega) = (\ell^\infty:\omega)$ and A belongs to these spaces iff the rows of A belong to ℓ.*

5B. EXAMPLE. From Example 5, *$A \in (cs:\ell^\infty)$ iff* $A^T \in (\ell:bv)$. This may be read from Example 1C: *$\sup_n \sum_k |a_{nk}-a_{n,k-1}| < \infty$. For $(cs:c_o)$ and $(cs:c)$ add* (8.3.6) *that the columns of A are null, convergent, respectively.* From 8.3.13 $A \in (cs:Y)$ for some Y iff the rows of A belong to bv and *this is the condition for $A \in (cs:\omega)$.*

5C. EXAMPLE. From Example 5, *$A \in (bs:\ell^\infty)$ iff* $A^T \in (\ell:bv_o)$. This may be read from Example 1C: *the condition of Example 5B, and each row of A is null.* From 8.3.13, *the rows of A belong to bv_o and this is the condition for $A \in (bs:\omega)$.*

5D. EXAMPLE. From Example 5, *$A \in (\ell^p:\ell^\infty)$, $p > 1$,* iff $A^T \in (\ell:\ell^q)$, $1/p + 1/q = 1$. This may be read from Example 1D: *$\sup_n \sum_k |a_{nk}|^q < \infty$. For $(\ell^p:c_o)$ and $(\ell^p:c)$ add* (8.3.6) *that the columns of A are null, convergent respectively.* From 8.3.13, *the rows of A belong to ℓ^q and this is the condition for $A \in (\ell^p:\omega)$.*

6. EXAMPLE. Take $Y = bs$ in 8.3.10. For X any one of our nine spaces, *$(X:bs) = (X^{\beta\beta}:bs)$ and $A \in (X:bs)$ if $A^T \in (bv_o:X^\beta)$.*

6A. EXAMPLE. From Example 6, *$(c_o:bs) = (c:bs) = (\ell^\infty:bs)$ and A belongs to these spaces* iff $A^T \in (bv_o:\ell)$. This may be read from Example 2D: *$\sup_m \sum_k |\sum_{n=1}^m a_{nk}| < \infty$. For $(c_o:cs)$ add* (8.3.6) *that the column-sums of A are convergent; for $(c:cs)$ add to this* (8.3.7) *that $\sum_n \sum_k a_{nk}$ is convergent.*

The fact that $(c_o:cs) \subseteq \Phi$ is not at all obvious from this result. However it is immediate from the inclusion $(c_o:cs) \subset (c_o:c_o)$.

6B. EXAMPLE. From Example 6, *$A \in (cs:bs)$ iff* $A^T \in (bv_o:bv)$. This may be read from Example 2C: *$\sup_m \sum_k |\sum_{n=1}^m a_{nk} - a_{n,k-1}| < \infty$. For $(cs:cs)$ add* (8.3.6) *$\sum_n a_{nk}$ is convergent for each k.* This is the historic series-series map. *The conditions for regularity* i.e. $\sum (Ax)_n = \sum x_n$ *are $A \in (cs:cs)$ and $\sum_n a_{nk} = 1$ for each k.* To see this note that cs is an AK space and so, with $f(x) = \sum (Ax)_n$ we have $f(x) = \sum f(\delta^k)x_k = \sum_k \sum_n a_{nk}x_k$, giving sufficiency. For necessity take $x = \delta^k$.

6C. EXAMPLE. From Example 5, *$A \in (bs:bs)$ iff* $A^T \in (bv_o:bv_o)$. This may be read from Example 2C: *the condition of Example 6B and each row of A is a null sequence.*

6D. EXAMPLE. From Example 6, *$A \in (\ell^p:bs)$, $p > 1$, iff* $A^T \in (bv:\ell^q)$, $1/p + 1/q = 1$. This may be read from Example 2D: *$\sup_m \sum_k |\sum_{n=1}^m a_{nk}|^q < \infty$. For $(\ell^p:cs)$ add* (8.3.6) *$\sum_n a_{nk}$ is convergent for each k.*

7. EXAMPLE. Take $Y = bv$ in 8.3.10. For X any one of our nine spaces, *$(X:bv) = (X^{\beta\beta}:bv)$ and $A \in (X:bv)$ iff $A^T \in (cs:X^\beta)$.*

7A. EXAMPLE. From Example 7, *$(c_0:bv) = (c:bv) = (\ell^\infty:bv)$.* The condition is given in Example 3A. We shall show that *$(c_0:bv_0) = (c:bv_0) = (\ell^\infty:bv_0)$* as well, and this condition can also be found in Example 3A. The argument is this: $(c_0:bv_0) \subset (c_0:bv) = (\ell^\infty:bv) \subset (\ell^\infty:c)$ from Example 7; also $(c_0:bv_0) \subset (c_0:c_0)$ so $(c_0:bv_0) \subset (\ell^\infty:c) \cap (c_0:c_0) = (\ell^\infty:c_0)$ by 1.7.19. From this $(c_0:bv_0) \subset (\ell^\infty:bv) \cap (\ell^\infty:c_0) = (\ell^\infty:bv_0)$. The opposite inclusion is trivial.

7B. EXAMPLE. From Example 7 with A, A^T interchanged, we have *$A \in (cs:\ell)$ iff* $A^T \in (c_0:bv)$. This may be read from Example 3A:

$$\sup\{\sum_k \Big|\sum_{n\in K} a_{nk}-a_{n,k-1}\Big| : K \text{ a finite set of positive integers}\} < \infty.$$

8. EXAMPLE. Take $Y = \ell^p$, $p > 1$, in 8.3.10. For X any one of our nine spaces, *$(X:\ell^p) = (X^{\beta\beta}:\ell^p)$* and *$A \in (X:\ell^p)$ iff $A^T \in (\ell^q:X^\beta)$, $1/p + 1/q = 1$.*

8A. EXAMPLE. From Example 8, $(c_0:\ell^p) = (c:\ell^p) = (\ell^\infty:\ell^p)$, $p > 1$. The condition is given in Example 3B.

8B. EXAMPLE. From Example 8 with A, A^T and p, q interchanged, we have *$A \in (\ell^p:\ell)$, $p > 1$,* iff $A^T \in (c_0:\ell^q)$, $1/p + 1/q = 1$. This may be read from Example 3B:

$$\sup\{\sum_k \Big|\sum_{n\in N} a_{nk}\Big|^q : N \text{ a finite set of positive integers}\} < \infty.$$

9. EXAMPLE. Take $Y = \ell$ in 8.3.10. For X any one of our nine spaces, *$(X:\ell) = (X^{\beta\beta}:\ell)$ and $A \in (X:\ell)$ if and only if $A^T \in (c_0:X^\beta)$.* (Remark 8.3.11 cautions against using $(\ell^\infty:X^\beta)$ here.)

9A. EXAMPLE. From Example 9, $(c_o:\ell) = (c:\ell) = (\ell^\infty:\ell)$. The condition is given in Example 3B. Note that 8.3.8 implies that if $A \in (\ell^\infty:\ell)$ then $A^T \in (\ell^\infty:\ell)$ also. The condition given in Example 3B does not obviously have this symmetry property.

9B. EXAMPLE. From Example 9, *$A \in (bs:\ell)$ if and only if* $A^T \in (c_o:bv_o)$. The condition may be read from Example 3A: *the condition of Example 7B and each row of A is null.*

8.5. MAPPING CONCLUDED

For this section only we introduce a new space. It does not include ϕ.

1. DEFINITION. $cs_o = \{x \in cs: \sum x_i = 0\}$.

2. Clearly $cs = cs_o \oplus \delta^1$. Hence $cs_o^\beta = cs^\beta = bv$.

3. LEMMA. *Let A be a matrix and $b_{nk} = a_{nk} - a_{n,k+1}$. Let x be a sequence and $v_k = \sum_{i=1}^{k} x_i$.*

(i) If the rows of A are null sequences and $x \in bs$, or

(ii) If the rows of A are bounded sequences and $x \in cs_o$,

then $Ax = Bv$ and $x \in \omega_A$ if $v \in \omega_B$.

By Abel's identity (1.29) we have

$$\sum_{k=1}^{m-1} a_{nk}x_k = \sum_{k=1}^{m-1} a_{nk}(v_k - v_{k-1}) = \sum_{k=1}^{m-1} b_{nk}v_k + a_{nm}v_{m-1}.$$

In case (i) v is bounded and $a_{nm} \to 0$ as $m \to \infty$; in case (ii) $v_m \to 0$ and $\{a_{nm}\}$ is bounded for each n. Letting $m \to \infty$ yields the result.

4. LEMMA. *Let Y be an FK space. Then $A \in (cs_o:Y)$ iff it has convergent rows and $B \in (c_o:Y)$ where B is defined in Lemma 3.*

Necessity: Since $\omega_A \supset cs_o$ it follows from Remark 2 that $\omega_A \supset cs$ and so the rows of A are in $cs^\beta = bv \subset c$. If $v \in c_o$ let $x_n = v_n - v_{n-1}$. Then $x \in cs_o$ so Ax is defined. By Lemma 3(ii) $Bv = Ax \in Y$.

Sufficiency: If $x \in cs_o$ define v as in Lemma 3. Then $v \in c_o$ so Bv is defined and by Lemma 3(ii), $Ax = Bv \in Y$.

5. EXAMPLE. *Let Y be an FK space. Then $A \in (cs:Y)$ iff it has convergent rows, its first column $\in Y$ and $B \in (c_o:Y)$ where B is defined in Lemma 3.* Necessity: The first column of A is $A\delta^1 \in Y$. The rest is by Lemma 4. Sufficiency: By Lemma 4 and Remark 2.

5A. EXAMPLE. *$A \in (cs:\ell^p)$, $p \geq 1$, iff*

(*) $\quad sup\{\sum_n |\sum_{k\in K}(a_{nk}-a_{n,k-1})|^p : K$ *a finite set of positive integers*$\} < \infty$.

Necessity: From Example 5 and 8.4.3B we get (*) with $k-1$ replaced by $k+1$. Thus (*) is correct if K is forbidden to contain 1 since we can substitute $m = k+1$ and change the signs. By Example 5 the first column $\in \ell^p$ and this extends the condition to allow K to contain 1 again.

Sufficiency: Taking $K = \{1\}$ in (*) shows that the first column of $A \in \ell^p$. Condition (*) implies the existence of M such that $|\sum_{k\in K}(a_{nk}-a_{n,k-1})|^p < M$ for all K, n. Fix n, let $b_k = a_{nk} - a_n$ and take the pth root. Then $|\sum_{k\in K} b_k| < C$. Fix an arbitrary positive integer m and let $K_1 = \{k:k \leq m,\ b_k \geq 0\}$, $K_2 = \{k:k \leq m,\ b_k < 0\}$. Then $\sum_{k=1}^m |b_k| = \sum\{b_k:k \in K_1\} - \sum\{b_k:k \in K_2\} < 2C$. Hence the nth row of $A \in bv \subset c$. The result now follows by Example 5 and 8.4.3.

5B. EXAMPLE. Imitating Example 5A, using 8.4.3A, we have *$A \in (cs:bv)$ iff*

$$\sup\{\sum_n |\sum_{k\in K}(a_{nk}-a_{n,k-1}-a_{n-1,k}+a_{n-1,k-1})| : K \text{ a finite set of positive integers}\} < \infty.$$

For $(cs:bv_o)$ add (8.3.6) *that the columns of A are null.*

6. EXAMPLE. *Let Y be an FK space. Then $A \in (bs:Y)$ iff it has null rows and $B \in (\ell^\infty:Y)$ where B is defined in Lemma 3.*

Necessity: Since $\omega_A \supset bs$ the rows of A are in $bs^\beta = bv_o \subset c_o$. If $v \in \ell^\infty$ let $x_n = v_n - v_{n-1}$. Then $x \in bs$ so Ax is defined. By Lemma 3(i) $Bv = Ax \in Y$.

Sufficiency: If $x \in bs$ define v as in Lemma 3. Then $v \in \ell^\infty$ so Bv is defined and by Lemma 3(i), $Ax = Bv \in Y$.

6A. EXAMPLE. $A \in (bs:bv)$ iff it has null rows and $B \in (c_o:bv)$ This is by Example 6 and 8.4.7A. From 8.4.3A, *$A \in (bs:bv)$ iff it has null rows and*

$$\sup\{\sum_n |\sum_{k\in K} a_{nk}-a_{n,k+1}-a_{n-1,k}+a_{n-1,k+1}| : K \text{ a finite set of positive integers}\} < \infty.$$

6B. EXAMPLE. $A \in (bs:bv_o)$ iff it has null rows and $B \in (c_o:bv_o)$. This is by Example 6 and 8.4.7A. We shall show that *$A \in (bs:bv_o)$ iff $A \in (bs:bv)$ (Example 6A) and A has null columns.*

Sufficiency: Clearly B has null columns and $B \in (c_o:bv)$ by Example 6A. Thus $B \in (c_o:bv_o)$ by 8.4.3A. So, as just remarked, $A \in (bs:bv_o)$.

Necessity: We have only to show that A has null columns. Now, taking $x = \delta^m$ in Lemma 3(i), $A\delta^m = Bv$ where v is a certain bounded sequence. Since $B \in (c_o:bv_o) = (\ell^\infty:bv_o)$ by 8.4.7A it follows that Bv is a null sequence, hence so is $A\delta^m$. This is the mth column of A.

6C. EXAMPLE. *$A \in (bs:\ell^p)$, $p \geq 1$ iff* it has null rows and $B \in (c_o:\ell^p)$. This by Example 6 and 8.4.8A. Thus by 8.4.3B the condition is: *A has null rows and*

$$\sup\{\sum_n |\sum_{k\in K} a_{nk}-a_{n,k+1}|^p : K \text{ a finite set of positive integers}\} < \infty.$$

6D. EXAMPLE. From Example 6, *$A \in (bs:c)$* iff it has null rows and $B \in (\ell^\infty:c)$ i.e. *iff A has null rows, convergent columns and $\sum_k |a_{nk}-a_{n,k+1}|$ is uniformly convergent.* Sufficiency is obvious by applying 1.7.18 to B. Conversely if $B \in (\ell^\infty:c)$ and A has null rows we obtain the fact that the columns of A are convergent by taking $x = \delta^m$ in Lemma 3(i); $A\delta^m = Bv$ where v is a certain bounded sequence so $A\delta^m \in c$.

6E. EXAMPLE. $A \in (bs:c_0)$ iff $A \in (bs:c)$ and A has null columns. The conditions may be read from Example 6D. The proof is the same as the preceding, using 1.7.19.

7. LEMMA. *Let A be a matrix, $b_{nk} = \sum_{i=1}^{n} a_{ik}$. Then $\omega_B = \omega_A$ and for $x \in \omega_A$, $(Bx)_n = \sum_{i=1}^{n} (Ax)_i$.*

For $x \in \omega_B$, $(Bx)_n - (Bx)_{n-1} = \sum_k (b_{nk}-b_{n-1,k})x_k = (Ax)_n$, hence $x \in \omega_A$ and the stated identity holds. Conversely if $x \in \omega_A$ it is trivial that $x \in \omega_B$ and again the identity holds.

8. EXAMPLE. *$A \in (\ell^\infty:cs)$ if and only if $B \in (\ell^\infty:c)$ where B is given in Lemma 7. This is clear from Lemma 7. Hence by 1.7.18(ii), $A \in (\ell^\infty:cs)$ iff $\sum_k |\sum_{i=n}^{\infty} a_{ik}| \to 0$.*

Because $(c_0:Y) = (\ell^\infty:Y)$ for several spaces Y, it is worthwhile to point out that this is false for $Y = c_0$ and $Y = cs$. The identity matrix lies in $(c_0:c_o)\setminus(\ell^\infty:c_0)$ while

$A \in (c_0:cs)\setminus(\ell^\infty:cs)$ where $(Ax)_n = x_n - x_{n+1}$.

9. EXAMPLE. From Examples 8 and 6, $A \in (bs:cs)$ iff the rows of A are null and $\sum_k |\sum_{i=n}^{\infty} a_{ik}-a_{i,k+1}| \to 0$.

10. EXAMPLE. From Example 8.4.7, *$A \in (\ell^p:bv)$, $p > 1$, iff* $A^T \in (cs:\ell^q)$, $1/p + 1/q = 1$. This may be read from Example 5A:

$$\sup\{\sum_k |\sum_{n\in N} (a_{nk}-a_{n-1,k})|^q : N \text{ a finite set of positive integers}\} < \infty.$$

For $(\ell^p:bv_0)$ add (8.3.6) that each column of A is null (as in 8.4.1C).

11. EXAMPLE. From 8.4.7, $A \in (bs:bv)$ if and only if $A^T \in (cs:bv_0)$. This may be read from Example 5B.

The remaining case is $(\ell^p:\ell^r)$, $p > 1$, $r > 1$. No tidy conditions are known. Elegant theorems of Crone and Pitt are given in [55] p. 114, Theorem 5.4 and p. 140, Theorem 9.2, respectively. In [38] an explicit norm is given for a Hausdorff matrix; in this special case the results are easier to apply than those of Crone. Other conditions are given in [72]. A very nice discussion of the Hilbert matrix, which lies in $(\ell^2:\ell^2)$, is given in [22B], p. 208, Solution V.

12. EXAMPLE. *Let A be a matrix, $B = |A|$ and assume that $B1$ and B^T1 are both bounded. Then $A \in (\ell^p:\ell^p)$ for $p > 1$.*

This is the special case $z = 1$ of:

13. EXAMPLE. *Let $1/p + 1/q = 1$ and suppose there exists a sequence z of positive numbers such that $(Bz^q)_n \le \alpha z_n^q$, $(B^Tz^p)_k \le \beta z_k^p$ for all n,k; α,β constant, $B = |A|$. Then $A \in (\ell^p:\ell^p)$.*

Let $x \in \phi$ with $x_k \geq 0$ for all k. Fix m and set

$$u = \sum_{n=1}^{m} |(Ax)_n|^p \leq \sum_{n=1}^{m} (\sum_k b_{nk}x_k)^p$$

$$= \sum [\sum_k (b_{nk}z_k^q)^{1/q}(b_{nk}z_k^{-p})^{1/p}x_k]^p$$

Applying Hölder's inequality (7.3.8) in which we take the pth power of each side: $u \leq$

$$\sum (\sum_k b_{nk}z_k^q)^{p/q} (\sum_k b_{nk}z_k^{-p}x_k^p) \leq \alpha^{p/q} \sum_{n=1}^{m} z_n^p \sum_k b_{nk}z_k^{-p}x_k^p$$

$$= \alpha^{p/q} \sum_k z_k^{-p}x_k^p \sum_{n=1}^{m} b_{nk}z_n^p \leq \alpha^{p/q} \beta \sum x_k.$$

This shows that $A \in (\phi:\ell^p)$ since the argument could have been written with $|x|$ instead of x; also, giving ϕ the ℓ^p norm it shows that $\|A\| \leq \alpha^{1/q}\beta^{1/q}$. Since ℓ^p has AK it follows that $A \in (\ell^p:\ell^p)$.

14. EXAMPLE. $(\ell:\ell) \cap (\ell^\infty:\ell^\infty) \subset (\ell^p:\ell^p)$ *for all* $p > 1$. *The inclusion is proper in the strong sense that* $(C,1)$ *belongs to the right side for all* $p > 1$, *but not to the left.* The first part is immediate from Example 12 with 8.4.1D, 8.4.5A. That $(C,1) \notin (\ell:\ell)$ was pointed out in 8.4.1D. Finally we shall prove that the criterion of Example 13 applies. Let $z_n = n^{-\varepsilon}$ with $0 < \varepsilon < 1/q$. Then, with $B = (C,1)$, $(Bz^q)_n = n^{-1} \sum_{k=1}^{n} k^{-\varepsilon q} < n^{-1} \int_0^n t^{-\varepsilon q}dt = z_n^q/(1-\varepsilon q)$; $(B^Tz^p)_k = \sum_{n=k}^{\infty} n^{-1}z^{-\varepsilon p} = \sum n^{-r}$ (where $r = 1+\varepsilon p$ so that $r > 1$). For $k = 1$ this is a certain (finite) number, while for $k > 1$ it is less than $\int_{k-1}^{\infty} t^{-r}dt = 1/[(r-1)(k-1)^{r-1}] \leq \beta k^{1-r} = \beta z_k^p$ where $\beta = 2^{r-1}/(r-1)$.

Some classical spaces are unions of FK spaces. Mapping theorems for these may be discussed in terms of maps of the constituent spaces. See [50].

8.6. FUNCTIONAL DUAL

There is an inclusion theorem of a much simpler and more easily applicable form than the ones in §8.2. It shows the condition, e.g. for c_o, in a more transparent form.

Recall the functional dual X^f given in 7.2.3 and the assumption, when it is used, that $X \supset \phi$.

1. THEOREM (A.K. Snyder and A. Wilansky). *Let X be an AD space. Then for any FK space Y we have $Y \supset X$ iff $Y^f \subset X^f$.*

Necessity is by 7.2.6. To prove sufficiency we apply 8.2.3. Let B be a subset of ϕ which is bounded in X. To show that B is bounded in Y it is sufficient to show it to be weakly bounded (8.0.2). To this end let $f \in Y'$. The hypothesis says that there exists $g \in X'$ such that $g(\delta^k) = f(\delta^k)$ for all k, hence $g = f$ on ϕ, in particular on B. Thus $f[B] = g[B]$ which is, by assumption, a bounded set of scalars.

2. EXAMPLE. *$Y \supset c_o$ iff $Y^f \subset c_o^f = \ell$ (7.3.1) i.e.* iff $\sum |f(\delta^k)| < \infty$ for all $f \in Y'$. Thus *$A \in (c_o:Y)$ iff $\{f(a^k)\} \in \ell$ for all $f \in Y'$.* The first condition is referred to as: $\{\delta^k\}$ is *weakly absolutely summable.* See [49] §1.2.

It is fairly easy to show directly the equivalence between this result and that of 8.2.7.

An easy remark is that *any weakly sequentially complete (e.g. reflexive) space $Y \supset c_o$ must also satisfy $Y \supset \ell^\infty$.* (Hence since ℓ^∞ is not weakly sequentially complete, the inclusion must be proper.) For $x \in \ell^\infty$ implies that $\sum x_k\delta^k$ is weakly Cauchy in Y, hence weakly convergent.

3. EXAMPLE. $Y \supset \ell$ iff $Y^f \subset \ell^f = \ell^\infty$, (7.3.1) i.e. iff $\{\delta^n\}$ is weakly bounded. Here "weakly" may be omitted (8.0.2) and so

the result of 8.2.5 is reached again. Also $A \in (\ell : Y)$ iff $\{g(a^k)\} \in \ell^\infty$ for all $g \in Y'$.

4. EXAMPLE. $Y \supset bv_0$ iff $Y^f \subset bv_0^f = bs$, (7.3.5). This condition is: for each $f \in Y'$ there exists M such that $M > |\sum_{k=1}^{m} f(\delta^k)| = f(1^{(m)})$, i.e. $\{1^{(m)}\}$ is (weakly) bounded. As in Example 3 we have the result of 8.2.6 again. Another important formulation is *$Y \supset bv_0$ if $\{f(\delta^n)\} \in bs$ for each $f \in Y'$.* Similar to this is the result *$A \in (bv_0 : Y)$ iff (the columns of A belong to Y) and $\{g(a^k)\} \in bs$ for all $g \in Y'$.* This is from 8.4.2. and the fact that weakly bounded is the same as bounded. (8.0.2).

5. EXAMPLE. $Y \supset cs$ if and only if $Y^f \subset cs^f = bv$, (8.3.5). This condition is: for each $f \in Y'$, $\sum|f(\delta^k) - f(\delta^{k-1})| < \infty$. (Convention: $\delta^0 = 0$.) i.e. $\sum |f(d^k)| < \infty$ where $d^k = \delta^k - \delta^{k-1}$.

6. EXAMPLE. *$Y \supset \ell^p$ $(p > 1)$ iff $\{\delta^k\}$ is weakly absolutely q-summable $(p^{-1}+q^{-1} = 1)$ in Y.* This means that $\sum|f(\delta^k)|^q < \infty$ for all $f \in Y'$. This follows from Theorem 1 and $(\ell^p)^f = \ell^q$, Example 7.3.7. Also $A \in (\ell^p : Y)$ iff $\{g(a^k)\} \in \ell^q$ for all $g \in Y'$.

7. EXAMPLE. *Theorem 1 fails for β-duals.* In 10.3.23, 10.3.24 it is shown that there exists a space X and an AK space Y with $Y \not\subset X$, $Y^\beta = X^\beta$.

8. In contrast with Example 7, Theorem 1 implies this dual characterization theorem for f-duals: *Two AD spaces with the same f-dual are equal.* Of course $c^f = c_0^f = \ell$ but $c \neq c_0$. Hence *two AK spaces with the same β-dual are equal.* (7.2.7). An amusing corollary of Theorem 1 is that c_0 is the largest AD space X such that $X^{\beta\beta} = \ell^\infty$; indeed *if X is an AD space and*

Y an AK space with $Y^\beta = X^\beta$ *then* $Y \supset X$ since $Y^f = Y^\beta = X^\beta \subset X^f$ by 7.2.7.

A few more remarks: *if* $\phi \subset Y \neq \omega$ *there must exist* $f \in Y'$ *with* $f(\delta^k) \neq 0$ *for infinitely many* k for if not, $Y^f = \phi$ and so by Theorem 1, $Y = \omega$. Finally, *there exists no* Y *such that* $Y^f = \omega$. For such a Y would, by Theorem 1, satisfy $Y_0 \subset Z$ for every FK space $Z \supset \phi$. (Here $Y_0 = c\ell_Y\phi$.) This implies $Y_0 = \phi$ [if $u \notin \phi$, let $u \cdot v \notin c_0$, then $u \notin v^{-1} \cdot c_0$] which is impossible by 4.0.5.

Conditions of the type given in 8.4.3B are intimately connected with unconditional convergence. (They are obviously implied by absolute convergence.) An interesting discussion of this point may be found in [47B], pp. 11-13.

CHAPTER 9
SEMICONSERVATIVE SPACES AND MATRICES

9.1 INTRODUCTION.

Because of the historical roots of summability in convergence, conservative spaces and matrices play a special role in its theory. However, the results seem mainly to depend on a weaker assumption, that the spaces be semiconservative. This is a significant generalization of the theory, but more than that — it seems to be a more natural home for it. Many of the results still to come were first given for conservative spaces. In addition to these we also extend results given earlier so that the reader may compare the settings. (See for example 9.2.10, 9.3.6, 9.3.7, 9.6.8, 9.6.9).

Even if none of the previous remarks is sufficient to motivate the concept, semiconservativity must arise in the discussion of distinguished spaces in view of 10.2.7.

9.2. SEMICONSERVATIVE SPACES

1. DEFINITION. *An FK space X is called semiconservative (= sc for short) if $X^f \subset cs$.*

This means that $X \supset \phi$ and $\sum f(\delta^k)$ is convergent for each $f \in X'$.

2. EXAMPLE. *Every conservative space is sc*, for X conservative implies $X^f \subset c^f = \ell \subset cs$ by 7.2.6 and 7.3.1. The same proof, with c replaced by c_o, shows that *if $X \supset c_o$ then X is sc.* The introduction of sc is motivated by the observation that in the

conservative theory, the convergence of $\sum f(\delta^k)$ was usually sufficient — absolute convergence being more than was needed.

3. EXAMPLE. *z^β is sc if and only if $z \in cs$.* Since z^β has AK (4.3.7), we have $z^{\beta f} = z^{\beta\beta}$ by 7.2.7 so z^β is sc if and only if $z^{\beta\beta} \subset cs$. This implies $z \in z^{\beta\beta} \subset cs$; conversely if $z \in cs$, $z^\beta \supset cs^\beta = bv$ and $z^{\beta\beta} \subset bv^\beta = cs$, using 7.3.5 and 7.2.2. As a special case *$cs = 1^\beta$ is not sc.*

Taking $z \in cs \backslash \ell$ gives an example, z^β, of a sc space which does not include c_o.

4. THEOREM. *(i) Every sc space $\supset bv_o$, (ii) the intersection of all sc spaces is bv_o, (iii) bv_o is not sc, (iv) there is no smallest sc space.*

(i) $X^f \subset cs \subset bs$ so $X \supset bv_o$ by 8.6.4.

(ii) The intersection $I \subset \cap\{z^\beta : z \in cs\} = cs^\beta = bv$ using Example 3 and 7.3.5. Also $I \subset c_o$ since c_o is sc by Example 2, so $I \subset bv \cap c_o = bv_o$. The opposite inclusion is by (i).

(iii). $bv_o^f = bs \not\subset cs$ by 7.3.5.

(iv). By (ii), (iii).

5. THEOREM. *The following are sc spaces: (i). A closed subspace $Y \supset \phi$ of a sc space X, (ii). Any FK space $Y \supset X$ where X is sc, (iii). $\cap X_n$ where each X_n, $n = 1,2,\ldots$, is sc.*

(i) is true since $Y^f = X^f$ (7.2.6). (ii) holds since $Y^f \subset X^f \subset cs$. To prove (iii): First, the intersection, X, is an FK space by 4.2.15. Every $f \in X'$ can be written $f = \sum_{k=1}^{m} g_k$ where each $g_k \in X_n'$ for some n. This is by 4.0.3, 4.0.8. Thus $\{f(\delta^n)\}$ is the sum of m sequences each of which is in cs.

6. EXAMPLE. *bv and bs are not sc* since their closed subspaces bv_o and cs are not. (Theorem 4, Example 3).

7. $X^\beta \subset cs$ is not sufficient for X to be sc since $bv^\beta = cs$. This is hardly surprising since this condition holds for every space containing 1.

8. For purposes of contrast we write three lists, each list consisting of a set of equivalent conditions. The first is: $X \supset bv_o$, $X^f \subset bs$, $\{1^{(n)}\}$ is (weakly) bounded in X (8.2.6). The second is: X is sc, $X^f \subset cs$, $1^{(n)}$ is *weakly Cauchy* i.e. $\{f(1^{(n)})\}$ is convergent for each $f \in X'$, equivalently $\sum f(\delta^k)$ is convergent. The third is: $X \supset c_o$, $X^f \subset \ell$, $\sum |f(\delta^k)| < \infty$ for each $f \in X'$. (8.6.2).

The extension of two-norm closure (§6.2) to sc spaces is made by using the bv norm:

9. LEMMA. *Let X be an FK space $\supset \phi$. Then $\|\cdot\|_{bv}$ is admissible for X. (6.2.1).*

Let $\|x\|_{bv}$ be written $\|x\|$. First, for $x \in bv$ we have $\|x^{(n)}\| \to \|x\|$ because $\|x^{(n)}\| = \sum_{k=1}^{n-1} |x_k - x_{k+1}| + |x_n|$. Also $x_n = \sum_{k=n}^{m-1} (x_k - x_{k+1}) + x_m$; combining these

$$\|x^{(n)}\| \leq \sum_{k=1}^{m-1} |x_k - x_{k+1}| + |x_m|.$$

Letting $m \to \infty$ gives $\|x^{(n)}\| < \|x\|$.

Next with $D = \{x \in X: \|x\| \leq 1\}$ and $D_n = \{x \in X: \|x^{(n)}\| \leq 1\}$, the result just proved shows that $D = \cap D_n$. Each D_n is closed since X is an FK space, so D is closed. The other part of the definition was part of the above proof.

10. THEOREM. *Let X be a sc space. Then $W \cap bv = 2_{bv}\phi = 2_{bv}\, bv_o$.*

By 6.2.4, 6.2.5, $W \cap bv \subset 2_{bv}\phi \subset 2_{bv}bv_0 \subset bv$. Finally let $z \in 2_{bv}bv_0$. It is sufficient to prove $z \in W$. To this end let $f \in X'$. Say $a^n \to z$, $a^n \in bv_0$, $\|a^n\|_{bv} < M$. Then

$$f(z) = \lim f(a^n) = \lim_n \sum_k a_k^n f(\delta^k) = \lim_A u \qquad (1)$$

where A is the matrix with $a_{nk} = a_k^n$ i.e. a^n is the nth row of A, and $u_k = f(\delta^k)$. This is true since $a \in bv_0$ implies $a = \sum a_k \delta^k$ in bv_0 (7.3.5) a fortiori, in X. Now $A \in (cs:c)$ from 8.4.5B for A has convergent columns, indeed $a_k^n \to z_k$ since X is an FK space, also $\sum_{k=1}^{m} |a_k^n - a_{k-1}^n| = |a_1^n| + \sum_{k=1}^{m-1} |a_k^n - a_{k+1}^n| \to |z_1| + \|a^n\|_{bv}$ as $m \to \infty$ since $a^n \in bv_0$. The last term is bounded by assumption and so the criteria of 8.4.5B are satisfied.

Since $A \in (cs:c)$, $\lim_A \in cs'$ (either by Banach-Steinhaus 1.0.4, or because $\lim_A = \lim \circ A$ with 4.2.8), also $u \in cs$ since X is sc, and since cs has AK, $\lim_A u = \lim_A(\sum u_k \delta^k) = \sum u_k \lim_A \delta^k = \sum f(\delta^k) \lim_n a_k^n = \sum f(\delta^k) z_k$. With (1) this shows that $z \in W$.

11. Naturally Theorem 10 holds if X is conservative or $X \supset c_0$. In this case $W \cap \ell^\infty = 2_\infty \phi$ as well, by 6.2.6. In [69] p. 601, it is shown that the latter result fails if X is only assumed sc, even if $X = c_A$. In the same place an example is given of $X \supset bv$ such that the conclusion of Theorem 10 fails. It is also shown that $1 \in 2_\infty \phi$ in c_A, A sc does not imply that A is conull. Some further examples and comments may be found in [68], especially 4.12.

9.3. COREGULAR AND CONULL.

1. DEFINITION. *A sc space is called conull if* $1^{(n)} \to 1$ *weakly.*

Here $1^{(n)} = \sum_{k=1}^{n} \delta^k$. The definition is the same as that of conull for conservative spaces, 4.6.2. A sc space need not contain 1, but must, by definition, if it is conull.

2. DEFINITION. *A sc space containing* 1 *is called coregular if it is not conull.*

Thus c_o is sc but is not classified as coregular. This is an arbitrary convention. Any sc space X not containing 1 is a closed subspace of $X \oplus 1$ (4.5.5) and the latter space is coregular. (X is convex closed hence weakly closed (5.0.2).) Thus it would do no harm to define a coregular space as a sc space which is not conull.

3. Henceforth a coregular or conull space will be assumed to be sc and to contain 1. Thus by 9.2.4 it must include bv.

FK spaces which include bv are called *variational* by J. J. Sember, who studied their properties. They need not be sc e.g. bv is not sc. (9.2.6.)

4. DEFINITION. *An FK space is called variational semiconservative (= vsc for short) if it is sc and includes bv.*

Thus $X \supset bv$ iff $X^f \subset bs$ and $1 \in X$, (8.6.4), while X is vsc iff $X^f \subset cs$ and $1 \in X$, also iff X is sc and $1 \in X$. (9.2.4(i).)

5. A conull space is automatically vsc while a coregular space is defined to be a vsc space which is not conull. *Every coregular or conull space* $\supset bv$ by Definition 4.

Some of the conservative theory extends very easily:

6. THEOREM. *Let X, Y be FK spaces with $X \subset Y$. If X is conull so is, Y. If X is coregular and closed in Y then Y is coregular.*

The proof 4.6.3 applies, taking account of 9.2.5.

7. THEOREM. *The intersection of countably many conull spaces is conull.*

By the proof of 4.6.6 taking account of 9.2.5.

Since X is conull iff $1 \in W$, we have X is conull iff $1 \in 2_{bv}\phi$ by 9.2.10. If X is conservative this is equivalent to the formally weaker condition $1 \in 2_\infty bv$ by 6.2.10. The first example cited in 9.2.11 shows that this equivalence fails for sc X in general.

8. THEOREM. *If bv_o is closed in a vsc space X, then X is coregular.*

Compare 4.6.4. Since $1^{(n)} \in bv_o$ and bv_o is weakly closed (5.0.2) it follows that $1^{(n)} \not\to 1$.

9. It is equivalent to assume in Theorem 8 that bv is closed. This is by 4.5.3.

Recall the functional χ defined on X' by $\chi(f) = f(1) - \sum f(\delta^k)$ where X is an FK space containing the sequences mentioned e.g. a vsc space.

10. THEOREM. *Let X be a vsc space. Then X is conull iff $\chi = 0$ on X'.*

This is precisely the content of Definition 1. (See Remark 5).

11. APPLICATION. *Suppose given a sequence $\{a_n\}$ of members of cs. Then there exists an unbounded sequence x such that $\sum a_{nk}x_k$ is convergent for each n.* Let $X = \cap a_n^\beta$. Then X is

conull by 9.2.3 and Theorem 7. (Each a_n^β is conull since it has AK (4.3.7) and contains 1.) Thus $X \not\subset \ell^\infty$ by Theorem 6. (The latter space is coregular by 4.6.4.)

9.4. MATRIX DOMAINS

The problem of characterizing matrices A such that Y_A is conservative (for a given Y) was solved in 8.4.3. In this section we solve the same problem for sc.

If A is a triangle, $f \in Y_A'$ if and only if $f = g \circ A$, $g \in Y'$, since $A: Y_A \to Y$ is an equivalence. Thus $f(\delta^k) = g(a^k)$ where a^k is the kth column of A, and $Y_A^f \subset cs$ iff $\sum g(a^k)$ converges for $g \in Y'$. This turns out to be the correct condition in general.

1. THEOREM. *Let Y be an FK space and A a matrix. Then Y_A is sc iff the columns of A are in Y and $\{g(a^k)\} \in cs$ for each $g \in Y'$, where a^k is the kth column of A, $a_n^k = a_{nk}$.*

Necessity: The columns of A are in Y since $Y_A \supset \phi$ by definition of sc. Given g, let $f(x) = g(Ax)$ for $x \in Y_A$, so $f \in Y_A'$ by 4.4.2. Then $f(\delta^k) = g(a^k)$ and the result follows since $Y_A^f \subset cs$.

Sufficiency: We first note that each row of A belongs to cs since in the hypothesis we may take $g = P_n$ where $P_n(x) = x_n$; this yields $\{g(a^k)\} = \{a_{nk}\}$. Hence $\omega_A \supset bv$.

Now let $f \in Y_A'$. Then by 4.4.2, $f(x) = \alpha x + g(Ax)$ with $g \in Y'$, $\alpha \in \omega_A^\beta \subset bv^\beta = cs$. Thus $f(\delta^k) = \alpha_k + g(a^k)$; by the hypothesis and the fact that $\alpha \in cs$ we have $\{f(\delta^k)\} \in cs$. Thus $Y_A^f \subset cs$ and Y_A is sc.

2. *Given A, if there is any Y such that Y_A is sc, then the rows of A lie in cs.* For $\omega_A \supset bv$ as proved in Theorem 1. Alternatively, one can argue that if r is a row of A,

$r^\beta \supset \omega_A \supset Y_A$ hence r^β is sc by 9.2.5(ii) and so $r \in cs$ by 9.2.3. Indeed, just as in 8.3.12, we have *given A, there exists Y such that Y_A is sc if and only if the rows of $A \in cs$.* Necessity was just proved. For sufficiency take $Y = \omega$; then $\omega_A = \cap\{r^\beta : r$ is a row of $A\}$ is sc by 9.2.3 and 9.2.5(iii). Another formulation: *ω_A is sc iff $\omega_A \supset bv$.*

3. Theorem 1 says Y_A is sc if $\{\sum a^k\}$ is weakly Cauchy in Y. (Compare 9.2.8.) Also 8.4.2 says that $Y_A \supset bv_0$ iff $\{\sum a^k\}$ is (weakly) bounded in Y. The two conditions can also be written: $\{g(a^k)\} \in cs, bs$, respectively, for all $g \in Y'$

Remark 2 says that the columns of A^T lie in cs. The next result improves this:

4. THEOREM. *If Y_A is sc then $A^T \in (Y^\beta : cs)$.*

First $A \in (bv_0 : Y)$ by 9.2.4. Hence $A^T \in (Y^\beta : bs)$ (8.3.8). Under stringent assumptions we could now apply the improvement of mapping in 8.3.6 since the rows of A are in cs. To prove the result in general, let $z \in Y^\beta$ and define $g \in Y'$ by $g(y) = zy$ using the Banach-Steinhaus theorem (1.0.4). Let $f(x) = g(Ax)$ so that $f \in Y'_A$ by 4.4.2. Hence $\{f(\delta^k)\} \in cs$. But $f(\delta^k) = g(A\delta^k)$ $= \sum z_n a_{nk} = (A^T z)_k$ so $A^T z \in cs$.

5. COROLLARY. *Let Y be a BK space and suppose that Y_A is sc. Then $A \in (bv : Y^{\beta f})$.*

By Theorem 4 and 8.3.8.

6. EXAMPLE. Let $Y = bv$, $A = I$. Then $A \in (bv:bv) = (bv:bv^{\beta f})$ but $Y_A = bv$ is not sc (9.2.6). Thus *the converse of Corollary 5 is false.* Also $A^T = I \in (cs:cs) = (Y^\beta : cs)$ so the *converse of Theorem 4 is false.* In this example *the columns of $A \in Y$.*

We can obtain a converse for Theorem 4 in the unimportant case in which Y has AK:

7. THEOREM. *Let Y be an FK space with AK. Then Y_A is sc iff the columns of A belong to Y and $A^T \in (Y^\beta : cs)$.*

Sufficiency: Apply Theorem 1. Let $g \in Y'$, $z_n = g(\delta^n)$. Then $z \in Y^f = Y^\beta$ by 7.2.7, so $A^T z \in cs$. But $(A^T z)_k = \sum z_n a_{nk}$ $= g(\sum a_{nk}\delta^n) = g(a^k)$ since Y has AK.

8. One can now apply Theorems 1 and 7 to give a list of conditions equivalent to the semiconservativity of Y_A for Y running through the spaces in the Table p. 116. This was done for $Y = \omega$ in Remark 2. For $Y = c$ it is 9.5.6.

9. THEOREM. *Let Y be an FK space and A a matrix such that Y_A is vsc. Then Y_A is conull iff $\sum g(a^k) = g(A1)$ for all $g \in Y'$ where a^k is the kth column of A. If not conull it is coregular.*

Necessity: Let $f(x) = g(Ax)$ so $f \in Y_A'$ by 4.4.2. Then $g(A1) = f(1) = \lim f(1^{(m)}) = \lim g(A1^{(m)}) = \sum g(a^k)$.

Sufficiency: Let $f \in Y_A'$. By 4.4.2 there are two cases to consider. First, $f(x) = \alpha x$, $\alpha \in \omega_A^\beta \subset bv^\beta = cs$ [$Y_A \supset bv$ by 9.2.4]. So $f(1-1^{(m)}) = \sum_{m+1}^{\infty} \alpha_k \to 0$. Second, $f(x) = g(Ax)$ for which the calculation given in the first part shows $f(1^{(m)}) \to f(1)$.

If Y_A is not conull we are allowed to call it coregular by 9.3.3.

9.5. MATRICES.

The historical order has now been reversed. First came conservative matrices, those for which $c_A \supset c$. When attention widened to FK spaces it was very natural to define one to be conservative if it includes c. But now we have semiconservative (sc) FK spaces and are contemplating what class of matrices might deserve the title. (This trend continues in §9.6, in which coregular and conull matrices are introduced after the corresponding FK spaces in §9.3.)

By analogy with the fact that A is conservative iff $A \in (c:c)$, it seems reasonable to look for an FK space X such that X_A is semiconservative iff $A \in (X:X)$. In this section we give a proof (due to A.K. Snyder) that no such X exists. The space X, to be sought in vain, must satisfy two conditions and we discuss them separately in Lemmas 1 and 2.

1. LEMMA. *The following are equivalent for an FK space X. (i). If $A \in (X:X)$ then X_A is sc (ii). X is sc.*

(i) implies (ii): Take $A = I$. (ii) implies (i): If $A \in (X:X)$ then $X_A \supset X$, hence X_A is sc by 9.2.5.

2. LEMMA. *Suppose that an FK space X has the property (i)'. If X_A is sc then $A \in (X:X)$. Then $X \subset bv$.*

If the conclusion is false, $X^\beta \not\supset cs$. [$X^\beta \supset cs$ implies $X \subset X^{\beta\beta} \subset cs^\beta = bv$ by 7.2.3 and 7.3.5.] Let $u \in cs \backslash X^\beta$, $0 \neq v \in X$, and let $a_{nk} = v_n u_k$. Then $X_A = \omega_A = u^\beta$ which is sc by 9.2.3; but $A \notin (X,X)$ since $\omega_A \not\supset X$.

3. COROLLARY. *There exists no FK space X such that X_A is sc iff $A \in (X:X)$.*

By Lemmas 1, 2 and 9.2.5, bv would be sc, contradicting 9.2.6.

However J.J. Sember [60] has proved that bv_A is sc iff $A \in (bv:bv)$ and $A: bv \to bv$ is a compact map. See also [10], Theorems 8, 9, 20.

It is amusing and a little unexpected that bv satisfies (i)' (Lemma 2). This follows from 9.4.5 with $Y = bv$. Thus bv is the largest space satisfying (i)'. Also any $X \subset bv_0$ must satisfy (i)' by 9.2.4(i). This leads to the conjecture that the converse of Lemma 2 is true. I have not been able to settle this.

4. DEFINITION. *A matrix A is called (variational) semi-conservative (=(v)sc for short) if c_A is (variational) semi-conservative.*

The reason for this definition is that summability theory deals with spaces of the form c_A and with FK spaces whose properties generalize those of such spaces. If we can extend theorems about conservative spaces to sc spaces so much the better.

5. EXAMPLE. *A may be sc and A^2 not.* Let $u \in c_0 \backslash bv$, $v \in cs \backslash u^\beta$, $a_{1k} = v_k$, $a_{n1} = u_n$, $a_{nk} = 0$ if $n > 1$ or $k > 1$. Then $c_A = v^\beta$ is sc by 9.2.3 but A^2 does not exist. Indeed $A(A\delta^1)$ does not exist so no natural modification of Corollary 3 will change its conclusion; e.g. there is no X such that A is sc iff $A \in (X:X)$.

6. THEOREM. *A is sc iff*

(i) A has convergent columns i.e. $c_A \supset \phi$,

(ii) $a \in cs$ *where* $a = \{a_k\}$, $a_k = \lim a_{nk}$,

(iii) $A^T \in (\ell:cs)$, *equivalently* $A \in (bv:\ell^\infty)$.

The equivalence of the conditions in (iii) is by 8.3.10.

Necessity: (i) is by definition; to prove (ii) apply 9.4.1 with $g = \lim$; (iii) is by 9.4.4. (The shorter proof mentioned there is available.)

Sufficiency: Apply 9.4.1. Let $g \in c'$ so that (1.0.2) $g(y) = \chi \lim y + ty$, $t \in \ell$. Then $g(a^k) = \chi a_k + (tA)_k$ and the result follows from (ii), (iii).

Theorem 6 provides a partial converse to 9.4.5, a modification of 9.4.7 for a space, c, without AK.

7. THEOREM. *A is sc if and only if Conditions (i), (ii) of Theorem 6 hold and*

(iii) $\sup_{m,n} \left| \sum_{k=1}^{m} a_{nk} \right| < \infty$ *(i.e.* $\ell_A^\infty \supset bv_o$*)*

(iv) the rows of A belong to cs (i.e. $\omega_A \supset bv$*).*

This is by Theorem 6 and 8.4.1B. The equivalences in parentheses are by 8.4.2A, 8.4.2, respectively.

8. COROLLARY. *A is sc if and only if* $c_A \supset bv_o$ *and Conditions (iv) (Theorem 7) and (ii) Theorem 6 hold.*

This is by Theorem 7 and 8.4.2A.

9. EXAMPLE. Let $(Ax)_n = \sum_{k=1}^{n} (x_{2k-1}-x_{2k})$. Then $c_A \supset bv$ either directly or by 8.4.2A. But *A is not sc* because $a_k = (-1)^{k-1}$ so (ii) fails in Theorem 6. The example can be modified to give a triangle by repeating each row and filling in the diagonal with a null sequence.

We add a few remarks in the spirit of Lemmas 1, 2.

10. Certain FK spaces X (all of which will be assumed to include ϕ in this paragraph) have the property

(*) *if* $X_A \supset bv$ *then* X_A *is sc.*

For example *every AK space satisfies (*)* for $A \in (X^\beta : cs)$ (8.3.8) and 9.4.7 gives the result. However *c does not satisfy (*)* by Example 9, hence ℓ^∞ *does not* by the following fact: *if Y satisfies (*) and X is a closed subspace of Y then X satisfies (*).* For $Y_A \supset X_A \supset bv$ implies Y_A is sc; X_A is a closed subspace (4.3.14), hence is sc by 9.2.5. *If* $X \supset bv$ *and* X *satisfies* (*) *then* X *is sc* — take $A = I$ — thus bv *does not satisfy* (*). I do not know any internal characterization of (*).

11. The converse property

(**) *If* X_A *is sc then* $X_A \supset bv$, *i.e.* X_A *is vsc*

is satisfied by every BK space with $X^{\beta f} = X$ (9.4.5). For example $X = \ell, bv, bs$. Another condition is given in 12.4.11.

9.6. COREGULAR AND CONULL MATRICES

To characterize vsc matrices we use again the notation $a_k = \lim a_{nk}$, $a = \{a_k\}$.

1. THEOREM. *A matrix* A *is vsc iff* $c_A \supset bv$ *and* $a \in cs$.

This is clear from 9.5.6.

2. Theorem 1 says that A *is vsc iff* $c_A \supset bv$ *and* $\chi(A)$ *exists.*

3. COROLLARY. *A matrix* A *is vsc iff (i)* A *has convergent columns, (ii)* $a \in cs$, *(iii)* $A1 \in c$, *(iv) The rows of* A *form a bounded set in* cs.

Necessity: (i), (ii) are the same as in 9.5.6. (iii) is implied by $c_A \supset bv$, (iv) is equivalent to (iii), (iv) in 9.5.7.

Sufficiency: (i), (iii) and (iv) imply $c_A \supset bv$ by 8.4.2A. The rest is by Theorem 1.

4. Compare Corollary 3 with the fact that A is conservative iff A has convergent columns, $A1 \in c$, and the rows of A form a bounded set in ℓ. The condition $a \in \ell$ corresponding to (ii) is true but need not be listed since it follows from the others

(1.3.7).

Before proceeding we need a result about associativity (partly due to R. DeVos).

5. LEMMA. *Let* A *be a matrix,* Z *an* AD *space, and* x *a sequence such that* (i). $(tA)x\ (=\sum_k \sum_n t_n a_{nk} x_k)$ *converges for each* $t \in Z$, *and* (ii). $t(Ax)\ (=\sum_n \sum_k)$ *converges for each* $t \in Z$. *Then the two sums are equal for each* $t \in Z$. *If* Z *is an* AK *space, condition* (ii) *and the conclusion are implied by* (i).

Example 13.4.16 will show that (even a greatly strengthened form of) (i) does not imply (ii) without the extra assumption. Example 12.3.4 will show that AD cannot be omitted. Let $f(t) = (tA)x$, $g(t) = t(Ax)$. Then $f, g \in Z'$ by the Banach-Steinhaus theorem (1.0.4). Further $f = g$ on ϕ (1.4.3) so $f = g$ on Z. If Z has AK, $f(t) = \sum t_i f(\delta^i) = \sum_i t_i \sum_k a_{ik} x_k = t(Ax)$.

Note that (i) says $A^T \in (Z:x^\beta)$. By 8.3.8, $A \in (x^{\beta\beta}:Z^f)$ which is a stronger statement than (ii) when Z has AK. (7.2.7).

6. In writing $(tA)x$ we treat t as a row vector. This can also be written $(A^T\mathbf{t})x$ in which t is a column vector.

7. LEMMA. *Let* Y *be a* BK *space such that* Y^β *has* AD. *Suppose that* $Y_A \supset bv$. *Then for* $t \in Y^\beta$, $x \in bv$ *we have* $(tA)x$ *exists and equals* $t(Ax)$.

Clearly $t(Ax)$ exists so it is sufficient by Lemma 5 to show the existence of $(tA)x$. Now $A \in (bv:Y)$ and so $A^T \in (Y^\beta:cs)$ by the second part of 8.3.8 hence $A^T t \in cs$. Since $x \in bv = cs^\beta$, $(A^T t)x$ exists. This is $(tA)x$. (Remark 6).

In Lemma 7, bv may be replaced by bv_o.

The next result is the extension of 4.4.7 to vsc matrices. Recall the representation $f(x) = \mu \lim_A x + t(Ax) + \alpha x$, $t \in \ell$, $\alpha \in c_A^\beta$ for $f \in c_A'$ given in 4.4.3; $\chi(f) = f(1) - \sum f(\delta^k)$ and $\chi(A) = \chi(\lim_A) = \lim_A 1 - \sum a_k$. Note that $\chi(A)$ exists if A is vsc by Remark 2.

8. THEOREM. *Let* A *be a vsc matrix,* $f \in c_A'$ *with the representation just given. Then* $\chi(f) = \mu\ \chi(A)$.

First $a = \{a_k\} \in cs$ by Theorem 1, $\alpha \in c_A^\beta \subset bv^\beta = cs$ so the proof of 4.4.7 applies, replacing ℓ by cs and using Lemma 7 with $Y = c$ for the last step.

As with conservative, vsc matrices A are divided into *coregular and conull according as* c_A *is coregular or conull.* (9.3.1, 9.3.2). Moreover the membership can be decided by the value of χ which was a matter of definition for conservative matrices, (1.6.1). The result is 4.6.1 turned around and generalized.

The proof now given for Theorem 9 short-cuts 9.4.9 by using only $g = \lim$ rather than all $g \in c'$, for $\chi(A) = 0$ is precisely the condition of 9.4.9 with $g = \lim$.

9. THEOREM. *A vsc matrix* A *is conull iff* $\chi(A) = 0$.

Let $\chi(A) = 0$. Then by Theorem 8, $\chi = 0$ on X'. This implies that X is conull by 9.3.10. Conversely if X is conull $\chi(\lim_A) = \chi(A) = 0$ by 9.3.10.

It should be emphasized that the generalization from conservative to vsc is as far as one can go with these ideas. For example if A is defined by $(Ax)_n = nx_n - nx_{n-1}$, there is no use in calling A conull just because $\chi(A) = 0$. Nothing follows. For example it is false that $1^{(n)} \to 1$ in c_A since this condition implies that A is sc by definition, but this A does not satisfy Condition (iii) of 9.5.7. To attempt to discuss variational matrices is equally futile since for such a matrix A, $\chi(A)$ does not even exist unless A is sc. (Remark 2).

10 COROLLARY. *A coregular matrix is* μ*-unique.* By Theorems 8, 9. This generalizes 4.4.8.

There are bigness theorems for c_A with A conull and sc namely $c_A \supset V$ where V is a certain oscillation space, (not Ω since it is conservative). However R. DeVos has scotched the FK program with an example. For details and extensive discussion see [22].

CHAPTER 10
DISTINGUISHED SUBSPACES OF FK SPACES

10.0. FUNCTIONAL ANALYSIS

1. Let Y be a quotient of X as in 7.0.1. Then the norm on Y is given by $\|y\| = \inf\{\|x\| : y = q(x)\}$. A set $E \subset Y$ is closed iff $q^{-1}[E]$ is closed in X. [80], Theorem 6-2-11.

2. Given an absolutely convex absorbing set U in a vector space, there exists a seminorm p such that $p(x) < 1$ implies $x \in U$ and the latter implies that $p(x) \leq 1$; p is called the *gauge* of U. [80], Theorem 2-2-1.

3. Examples of BK spaces X with $c_o \subset X \subset \ell^\infty$ with X not closed in ℓ^∞. These are all of the form $X = c_o + E$ where E is a closed subspace of ℓ^∞, hence a BK space. If $E \cap c_o = \{0\}$ as in (a), (b), (d), X is a BK space by 4.5.1; if not X is shown to be a BK space in [80], Example 13-4-5. (a) The easiest example is given in [82], p. 12, §6. (b) Let U be a subspace of ℓ which is not norming over c_o, [80], Example 15-4-7, $E = U^\perp$, [80] #15-4-31. (c) E = bs, [11], p. 29, Theorem 3. (d) E is a quasi-complement for c_o i.e. X is dense in ℓ^∞, [80], #15-3-301.

10.1. DUALS AS BK SPACES

Before launching the main topic in the next section we give some topological properties of duals of BK spaces. Duals in general are not FK spaces, for example, $\omega^\beta = \phi$ is not (4.0.5). For reference a few earlier results are collected:

1. THEOREM. *Let* X *be a* BK *space then* X^β, X^γ *and (assuming* $X \supset \phi$*)* X^f *are* BK *spaces.*

The first two are covered by 4.3.16, 4.3.17, the third by 7.2.14.

When any of these spaces is recognizable as a BK space, e.g. $c^f = \ell$, the norm is known and need not be calculated from Theorem 1. The calculation must yield an equivalent result by 4.2.4.

We know that $X^\beta \subset X^\gamma \subset X^f$ (7.2.7) and so the topologies are ordered in the opposite way. (4.2.4). Although it is not important, we take a moment to prove the stronger result $\|u\|_\beta = \|u\|_\gamma \geq \|u\|_f$ for $u \in X^\beta$. Let $f(x) = ux$ for $x \in X$. If $\|x\| \leq 1$ we have $|f(x)| = |ux| \leq \sup|\sum_{i=1}^n u_i x_i| = \|u \cdot x\|_{cs} \leq \|u\|_\beta$. Hence $\|f\| \leq \|u\|_\beta$. Since $f(\delta^n) = u_n$ the definition of the quotient norm (10.0.1) yields $\|f\| \geq \|u\|_f$. The next example shows that the norms need not be equivalent.

2. EXAMPLE. X^β *not closed in* X^f; *this implies that* X^γ *is not closed in* X^f. (4.3.18). Let z be an unbounded sequence and $X = c_0 \oplus z$ (4.5.5). Then $X^\beta = \ell \cap z^\beta$, $X^f = \ell$ (7.2.4) so $\phi \subset X^\beta \subset X^f = \ell$ and X^β is dense. More instructive considerations of this type are given in 14.5.4.

Criteria for X^β to be closed in X^f are given in 10.3.9, 14.5.3.

3. EXAMPLE. X^β *not closed in* X'. The embedding is given in 7.2.9. With X as in Example 2, $X' = \ell \oplus f$ for some f and the result is clear.

Criteria for X^β to be closed in X' are given in 10.3.11, 14.5.3.

The inclusion maps from X^β into X^γ and X^f (7.2.7(i)) are automatically continuous by 4.2.4. The same is true for X'. The embedding is shown in 7.2.9:

4. THEOREM. *Let X be a BK space. The embedding of X^β in X' is continuous.*

If $\|x\| \le 1$, $|\hat{u}(x)| = |ux| \le \sup|\sum_{k=1}^{m} u_k x_k| = \|u\cdot x\|_{cs} \le \|u\|_\beta$. Thus $\|\hat{u}\| \le \|u\|_\beta$. The result is also immediate from the diagram of 7.2.10.

A BK space with AK may be assumed to have a monotone norm (7.1.4). For some of the duals the norm is already monotone:

5. THEOREM. *Let X be a BK space. Then X^β and X^γ have monotone norms. (7.1.1).*

A direct check of these norms from 4.3.16, 4.3.17 shows this. For example $\|u\|_\beta = \sup\{\|u\cdot x\|_{cs} : \|x\| \le 1\}$. Now $\|u^{(m)}\cdot x\|_{cs} = \sup\{|\sum_{k=1}^{n} u_k x_k| : n \le m\}$. This is an increasing function of m. Further supposing that $\|u\|_\beta = 1$ there exists x with $\|x\| = 1$, $\|u\cdot x\|_{cs} > 1 - \varepsilon$. By definition this implies the existence of m such that $|\sum_{i=1}^{m} u_i x_i| > 1 - 2\varepsilon$, hence $\|u^{(m)}\cdot x\|_{cs} > 1 - 2\varepsilon$ which implies that $\|u^{(m)}\|_\beta > 1 - 2\varepsilon$ and yields $\sup\|u^{(m)}\| \ge 1$. Conversely for $x \in X$ and any m, $\|u^{(m)}\cdot x\|_{cs} \le \|u\cdot x\|_{cs}$; taking sup over $\|x\| \le 1$ yields $\|u^{(m)}\|_\beta \le \|u\|_\beta$. The only change for X^γ is to replace each cs by bs.

6. THEOREM. *Let X be a BK space $\supset \phi$, $x \in \phi$ and $u \in X^f$. Then $|ux| \le \|u\|\cdot\|x\|$. Also for each $x \in \phi$, $\|x\| = \sup\{|ux| : \|u\| \le 1\}$.*

For any $f \in X'$ with $f(\delta^k) = u_k$ we have $|ux| = |f(x)| \le \|f\|\cdot\|x\|$. Taking the inf over all such f yields the first result by definition of the norm on X^f. (10.0.1). To prove the second formula, let $f \in X'$ with $\|f\| = 1$, $f(x) = \|x\|$, using the Hahn-Banach theorem. Let $u_k = f(\delta^k)$. By definition of the quotient norm (10.0.1) $\|u\| \le 1$; also $|ux| = |f(x)| = \|x\|$. This, together with the first result proves the asserted equality.

A useful inclusion relation for multipliers follows from this. Let $M(X) = M(X,X)$, the latter symbol having been defined in 4.3.15.

7. LEMMA. *Let X be a BK space $\supset \phi$. Then $M(X) \subset M(X^f)$.*

Let $u \in M(X)$, $v \in X^f$. For $x \in \phi$ define $f(x) = \sum u_k v_k x_k$. Since $f(\delta^k) = u_k v_k$ it will follow that $u \cdot v \in X^f$ when it is shown that f is continuous on ϕ with the relative topology of X, since f can be extended to all of X by the Hahn-Banach theorem (3.0.1). This continuity holds since $|f(x)| \le \|v\| \cdot \|u \cdot x\|$ (Theorem 6) $\le \|v\| \cdot \|u\| \cdot \|x\|$ where $\|u\|$ is computed in $M(X)$ as in 4.3.15.

10.2. DISTINGUISHED SUBSPACES.

The subspaces given here have proved to be of value in discussing the fine structure of FK spaces and the properties of matrices. Early in the 20th century various classical results were associated with boundedness properties of matrices. (See §13.4). When placed in abstract form the property was seen to be equivalent to sectional boundedness, called AB for Abschnittsbeschrankte. (See §8.0 for bounded sets.)

1. DEFINITION. *An FK space $X \supset \phi$ is said to have AB if $\{x^{(n)}\}$ is a bounded set in X for each $x \in X$.*

It turned out that, for $X = c_A$, AB is equivalent to other properties such as $S \oplus 1 = F = B = X$ where S, F, B are some of the distinguished subspaces soon to be defined. The results referred to are given in 13.3.1. A natural process, then, is to consider spaces without AB and investigate the subspaces on their own merits. These investigations show the role of the subspaces in situations in which they do not satisfy the equations just given.

Examples to illustrate the possibilities will be given in later sections. An example of a non-AB space occurs in 7.2.5. There $\|u^{(m)}\| = \|u^{(m)}\|_\infty$ since $u^{(m)} \in c_o$. This is unbounded.

We begin with the smallest. The letter S stands for strong (convergence):

2. DEFINITION. *Let* X *be an FK space* $\supset \phi$. *Then* $S = S(X) = \{x: x^{(n)} \to x\} = \{x: x$ *has AK in* $X\} = \{x: x = \sum x_k\delta^k\}$. *If* A *is a matrix,* $S(A) = S(c_A)$.

Thus X is an AK space (4.2.13) iff $S = X$. For $X = c$ or ℓ^∞, $S = c_o$. Of course $S \supset \phi$ always. Also $S \subset X$ since X is complete.

The subspace $W = \{x \in X: f(x) = \sum x_k f(\delta^k)$ for all $f \in X'\}$ has been introduced earlier (5.6.1), and characterized in terms of two-norm convergence for conservative and vsc spaces. (6.2.6 ,9.2.10). It was also used to prove the Bounded Consistency Theorem 5.6.10. The letter W stands for weak (convergence).

3. DEFINITION. *Let* X *be an FK space* $\supset \phi$. *Then* $F^+ = F^+(X) = \{z \in \omega: \{z^{(n)}\}$ *is weakly Cauchy in* $X\} = \{z \in \omega: \{z_n f(\delta^n)\} \in cs$ *for all* $f \in X'\}$. *Also* $F = F^+ \cap X$. *If* A *is a matrix* $F(A) = F(c_A)$.

Thus $F \subset X$. If $X = c_o$, $1 \in F^+ \backslash F$. The letter F stands for functional (convergence) since $z \in F^+$ if and only if $\{f(z^{(n)})\}$ is convergent for all $f \in X'$. It is customary to write $z \in F^+$ as z has FAK i.e. functional AK. (Compare AK and SAK.)

4. DEFINITION. *Let* X *be an* FK *space* $\supset \phi$. *Then* $B^+ = B^+(X) = \{z \in \omega\colon \{z^{(n)}\}$ *is bounded in* $X\}$
$= \{z \in \omega\colon \{z_n f(\delta^n)\} \in bs$ *for all* $f \in X'\}$. *Also* $B = B^+ \cap X$. *If* A *is a matrix*, $B(A) = B(c_A)$.

Thus $B \subset X$. The two definitions are equivalent since $\sum_{n=1}^{m} z_n f(\delta^n) = f(z^{(m)})$, bounded and weakly bounded being the same (8.0.2.)

5. EXAMPLE. For $X = c$, $F^+ = B^+ = \ell^\infty$, $F = B = c$, $\overline{\phi} = S = W = c_0$.

By Definition 1, X is an AB space iff $B = X$. The space of 7.1.2 has $B^+ = \ell$, a proper subspace of X; $1 \notin B$ since $\|1^{(n)}\| = n$.

6. THEOREM. *Let* X *be an* FK *space* $\supset \phi$. *Then* $\phi \subset S \subset W \subset F \subset B \subset X$ *and* $\phi \subset S \subset W \subset \overline{\phi}$.

The only non-trivial part is $W \subset \overline{\phi}$. Let $f \in X'$, $f = 0$ on ϕ. A glance at the definition of W just given shows that $f = 0$ on W. The Hahn-Banach theorem (3.0.1) gives the result.

Example 5 shows that F need not be included in $\overline{\phi}$

These ideas are intimately bound with the ideas of Chapter 9: inclusions of bv_0, bv, sc and vsc (= variational semiconservative):

7. THEOREM. *Let* X *be an* FK *space* $\supset \phi$, $z \in \omega$, *then*

(i) $z \in B^+$ *iff* $z^{-1}\cdot X \supset bv_0$, *in particular* $1 \in B^+$ *iff* $X \supset bv_0$,

(ii) $z \in B$ *iff* $z^{-1}\cdot X \supset bv$, *in particular* $1 \in B$ *iff* $X \supset bv$,

(iii) $z \in F^+$ *iff* $z^{-1}\cdot X$ *is sc, in particular* $1 \in F^+$ *iff* X *is sc*, *(iv)* $z \in F$ *iff* $z^{-1}\cdot X$ *is vsc, in particular* $1 \in F$ *iff* X *is vsc,*

(v) $z \in W$ *iff* $z^{-1}\cdot X$ *is conull, in particular* $1 \in W$ *iff* X *is conull.*

(vi) $z \in S$ *iff* $z^{-1}\cdot X$ *is strongly conull; in particular* $1 \in S$ *iff* X *is strongly conull.*

Let $f \in (z^{-1}\cdot X)'$. By 4.4.10, $f(\delta^{(n)}) = \alpha_n + g(z\cdot\delta^n) = \alpha_n + g(z_n\delta^n) = \alpha_n + z_n g(\delta^n)$, $\alpha \in \phi$, $g \in X'$. Thus $\{f(\delta^n)\} \in bs, cs$ iff $z \in B^+, F^+$ respectively. Parts (i), (iii) follow from 8.6.4 and 9.2.1.

If $z \in B$ then $z \in X$ so $1 \in z^{-1}\cdot X$ and so by (i), $z^{-1}\cdot X \supset bv$. Conversely $1 \in bv$ so $1 \in z^{-1}\cdot X$ i.e. $z \in X$. By (i), $z \in B$.

Part (iv) follows from (iii) in the same way.

Each condition in Part (v) implies that $z \in X$ so if $f \in (z^{-1}\cdot X)'$, $f(1-1^{(n)}) = g(z\cdot 1 - z\cdot 1^{(n)}) = g(z-z^{(n)})$ as in the proof of (i) (iii). The result is immediate from 5.6.1, 9.3.1.

For (vi) use 4.3.6 to obtain the seminorms of $z^{-1}\cdot X$. First $p_n(1-1^{(m)}) = (1-1^{(m)})_n = 0$ for $m > n$ while $g[z\cdot(1-1^{(m)})] = g(z-z^{(m)})$. The result is clear from this.

The first three conditions of 5.2.13 are $S = X$, $1 \in S$, $1 \in W$; 6.5.2 is a special case of Theorem 7 (v).

8. Theorem 7 suggests definitions of "distinguished subspaces" at will by choosing the set of z such that $z^{-1}\cdot X$ includes c or c_0 or is conservative and conull etc. As one example (which will not be used later) let $C^+ = \{z: z^{-1}\cdot X \supset c_0\}$, $C = C^+ \cap X$. From 8.6.2 it follows that $C^+ = X^{f\alpha}$ (4.3.17) and one can improve 6.5.1 (replacing c by c_0 for convenience) as follows. Let $X \supset c_0$, then $z^{-1}\cdot X \supset c_0$ iff $z \in X^{f\alpha}$.

In view of Remark 8, Theorem 7 (iv) is seen to be the analogue of 6.5.1.

The subspaces mentioned in Remark 8 could well merit serious study — however the subspaces S, W, F, B arose independently of the results of Theorem 7 at an early stage of the development of summability through functional analysis. Semiconservative spaces also arose independently as a natural home for some of the conservative theory. Only later, when both studies were established, was the simple link given in Theorem 7 observed.

9. THEOREM. *The distinguished subspaces are monotone i.e. if* $X \subset Y$ *then* $E(X) \subset E(Y)$ *where* $E = S, W, F, F^+, B, B^+$. *This holds also for* $E = \overline{\phi}$ *i.e.* $cl_X\phi \subset cl_Y\phi$.

The inclusion map $i: X \to Y$ is continuous (4.2.4), so $x^{(n)} \to x$ in X implies the same in Y. This is the assertion for S. For W it follows from the fact that i is weakly continuous (4.0.11).

Now $z \in F^+, B^+$ iff $\{z_n f(\delta^n)\} \in cs, bs$ respectively for all $f \in X'$, hence for all $g \in Y'$ since $g|X \in X'$ (4.2.4). The result follows for F^+, B^+ and so for F, B. The last part is by 4.2.7. (The results for F^+, B^+ are also immediate from 10.3.4, 10.4.2.)

This section is concluded with an unusual theorem which will not be referred to again. By $Y \cdot Z$ we mean $\{y \cdot z: y \in Y, z \in Z\}$. It need not be a vector space.

10. THEOREM. *Let* Y *be a sc space and* Z *an* AD *space. Suppose that an* FK *space* $X \supset Y \cdot Z$. *Then* $X \supset Z$.

Let $z \in Z$. Then $z^{-1} \cdot X \supset Y$ so $z \in F^+$ by Theorem 7. Hence $Z \subset F^+ = X^{f\beta}$ (obvious from Definition 3). Thus

$$Z^f \supset Z^\beta \supset X^{f\beta\beta} \supset X^f$$

and the result follows by 8.6.1. (If $1 \in Y$, the result is trivial.)

Theorem 10 has obvious connections with the construction of barrelled subspaces. See [80], Theorem 15-2-7.

Source [86].

10.3. THE SUBSPACE B.

Although X may have various pathological properties we shall see that its subspace B is relatively well-behaved, for example it has AB (Corollary 14) includes the other distinguished subspaces as closed subsets and includes $\{\delta^n\}$ as a basic set (Theorem 19.) But see Remark 27. Of course an AB space will have these properties too.

We begin with a growth theorem and an application.

1. THEOREM. *Let X be an FK space $\supset \phi$. Then for each $z \in B^+$ and each continuous seminorm p on X we have $z_n = O(p(\delta^n)^{-1})$.*

For $|z_n|p(\delta^n) = p(z_n\delta^n) = p(z^{(n)}-z^{(n-1)}) \le p(z^{(n)}) + p(z^{(n-1)}) < M$ since a continuous seminorm is bounded on bounded sets (8.0.4).

This result does not necessarily give any information, for example if $X = \omega$, $p(\delta^n) = 0$ for sufficiently large n, no matter what p is.

2. EXAMPLE. *Let X be an FK space $\supset \phi$ and suppose that some neighborhood U of 0 excludes all δ^n* e.g. if X is a BK space and $\|\delta^n\| \ge \varepsilon > 0$; *then $B^+ \subset \ell^\infty$.* There exists a continuous seminorm p such that $p(\delta^n) \ge 1$ for all n, for example take p to be the gauge of U (10.0.2). Theorem 1 gives the result.

3. EXAMPLE. An AB space with the property of Example 2 must contain only bounded sequences.

We now discuss B^+, B in some detail and give characterizations of AB spaces. Often the discussion is restricted to BK spaces for convenience. Remark 26 discusses restoring generality.

4. THEOREM. *Let* X *be an FK space* $\supset \phi$. *Then* $B^+ = X^{f\gamma}$

By 10.2.4, $z \in B^+$ iff $z \cdot u \in bs$ for each $u \in X^f$. This is precisely the assertion.

This makes it easy to compute B^+ and B. The next result makes it even easier although there is a little less here than meets the eye — namely B will be different for different Y e.g. Y may be AB and X not, even if Y is closed in X (7.1.2).

5. THEOREM. *Let* X *be an FK space* $\supset \phi$. *Then* B^+ *is the same for all FK spaces* Y *between* $\overline{\phi}$ *and* X; *i.e.* $\overline{\phi} \subset Y \subset X$ *implies* $B^+(Y) = B^+(X)$. *(The closure of* ϕ *is calculated in* X*).*

First $B^+(\overline{\phi}) \subset B^+(Y) \subset B^+(X)$ by 10.2.9. The first and last are equal by Theorem 4 and 7.2.4.

In particular this holds if Y is closed in X (assuming $Y \supset \phi$). Under the conditions of Theorem 5 the closure of ϕ is the same whether calculated in X or in Y. (10.2.9).

6. EXAMPLE. *Suppose that* c_o *is a closed subspace of an FK space* X. *Then* $B^+ = \ell^\infty$. From Theorem 5, $B^+(X) = B^+(c_o) = c_o^{f\gamma} = \ell^\gamma = \ell^\infty$. All this is actually quite trivial since for each x, $x^{(n)} \in c_o$. *The same result holds if* c *is a closed subspace of* X.

7. EXAMPLE. *A BK space* X *with* $c \subset X \subset \ell^\infty$, X *not closed in* ℓ^∞. This turns out to be quite hard. Since it rests mainly on functional analysis prerequisites we shall merely refer to a list

of constructions in 10.0.3. The easiest one to say is $c + bs$, although it is far from the easiest to check!

8. THEOREM. *Let X be an FK space $\supset \phi$. Then X has AB iff $X^f \subset X^\gamma$ i.e. $X^f = X^\gamma$.*

Necessity: Using Theorem 4, $X \subset B^+ = X^{f\gamma}$. Hence $X^\gamma \supset X^{f\gamma\gamma} \supset X^f$. Sufficiency: $B^+ = X^{f\gamma} \supset X^{\gamma\gamma} \supset X$. The last part is because $X^\gamma \subset X^f$ always. (7.2.7).

9. COROLLARY. *Let X be a BK space $\supset \phi$. If X has AB then X^β is closed in X^f.*

By Theorem 8, since X^β is closed in X^γ. (4.3.18.)

Closure of X^β in X' can be evoked from Corollary 9 by means of a property of the quotient map. The embedding is given in 7.2.9.

10. Let X be a BK space $\supset \phi$. Then *X^β is closed in X^f iff $X^\beta + \phi^\perp$ is closed in X'. This condition implies that X^β is closed in X'.* For $X^\beta + \phi^\perp = q^{-1}[X^\beta]$, using the notation of 7.2.10. The result follows from 10.0.1. The second part is by 4.5.2 with $H = X'$, $Z = X^\beta + \phi^\perp$; X^β is a BH space by 10.1.4, 10.1.1.

11. COROLLARY (W.L.C. Sargent). *Let X be a BK space with AB. Then X^β is closed in X'.*

This is by Corollary 9 and Remark 10.

Corollaries 9, 11 and their converses hold in a very strong form for every c_A. (14.5.3).

Next we give a short discussion of monotone norms, (7.1.1), culminating in Corollary 15. Any space with a monotone norm is automatically a BK space $\supset \phi$. We have seen that any AK space,

β-dual or γ-dual has (or may be assumed to have) a monotone norm. (7.1.4, 10.1.5.)

The first result is trivial:

12. THEOREM. *Any space with a monotone norm has AB.*

We shall see in Example 16 that the converse is false.

13. THEOREM. *Let $(X, \|\cdot\|)$ be a BK space $\supset \phi$. Then B^+ is a BK space with monotone norm $\|z\|_{B^+} = \sup\|z^{(n)}\|$; B is a BK space with $\|x\|_B = \|x\|_{B^+} + \|x\|$. If X has AB, there exists M such that $\|x^{(m)}\| \leq M\|x\|$ for all m and all $x \in X$.*

They are BK spaces by two applications of 10.1.1 (using Theorem 4) and the intersection Theorem 4.2.15, the latter giving the norm on B. The norm on B^+ is monotone since this space is a γ-dual. (10.1.5); this will also be obvious when the asserted identity is proved:

$\leq$: Let $z \in B^+$, $u \in X^f$. Then $|\sum_{k=1}^{m} u_k z_k| \leq \|u\| \cdot \|z^{(m)}\|$ by 10.1.6. Hence $\|u \cdot z\|_{bs} \leq \|u\| \cdot \sup\|z^{(m)}\|$ and by definition of the γ-norm, 4.3.17, (using $B^+ = X^{f\gamma}$, Theorem 4), $\|z\|_{B^+} \leq \sup\|z^{(m)}\|$.

$\geq$: Let $z \in B^+$. For each m, $\|z^{(m)}\| = \sup\{\sum_{k=1}^{m} u_k z_k : u \in X^f, \|u\| \leq 1\}$ (10.1.6) $\leq \sup\{\|u \cdot z\|_{bs} : \|u\| \leq 1\} = \|z\|_{B^+}$.

If X has AB there are two easy ways to see the existence of M. First, apply uniform boundedness to the maps $x \to x^{(n)}$; second, one can argue that the B-norm and the X-norm are equivalent since X is a BK space with each (4.2.4), and the result follows from the form of the B-norm.

14. COROLLARY. *Let X be a BK space $\supset \phi$. Then B^+ and B are AB spaces.*

The first is by Theorems 12, 13 (or 4 and 10.1.5). As to the second, for $x \in B$, $\|x^{(m)}\|_B = \|x^{(m)}\|_{B^+} + \|x^{(m)}\|$ by Theorem 13.

15. COROLLARY. *Let* X *be a* BK *space* $\supset \phi$. *Then* X *(has* AB *and) is a closed subspace of* B^+ *iff* X *has a monotone norm.*

Necessity is by Theorem 13: X has the same topology as B^+ by 4.2.4 so its norm is equivalent to that of B^+. Conversely if the norm is monotone, for $x \in X$, $\|x\| = \sup\|x^{(n)}\| = \|x\|_{B^+}$ by Theorem 13, so X is closed in B^+.

This result applies to AK spaces (7.1.4) and β and γ duals. (10.1.5).

16. EXAMPLE. *An* AB *space* X *with no equivalent monotone norm. Thus* X *is a non-closed subspace of* B^+ *and* B *is not closed in* B^+. This is the space of Example 7. By Theorem 5, $B^+ = \ell^\infty$.

It appears that there are two kinds of AB spaces X, those with monotone norms (X is closed in B^+) and the other kind as shown in Example 16.

An amusing but perhaps not significant remark is that $B^+(B^+) = B^+$ ($\supset$: by Corollary 14; $\subset$: by Theorem 13). This says $X^{f\gamma f\gamma} = X^{f\gamma}$. Many such formulas could be found. The corresponding result for W is false (12.5.7).

To prove the next theorem we need a result about multipliers. Let $M(X) = M(X,X)$; the latter symbol having been defined in 4.3.15.

17. LEMMA. *Let* X *be a* BK *space* $\supset \phi$. *These are equivalent: (i)* X *has* AB, *(ii)* $M(X) \supset bv$, *(iii)* $1 \in B[M(X)]$.

(ii) = (iii): by 10.2.7 (ii); (i) implies (iii): Using the norm for $M(X)$ given in 4.3.15, $\|1^{(m)}\| = \sup\{\|x\cdot 1^{(m)}\| : \|x\| \le 1\} = \sup\{\|x^{(m)}\| : \|x\| \le 1\} < M$ by Theorem 13.

(ii) implies (i): Fix $x \in X$. The map $T: bv \to X$ given by $Tu = u \cdot x$ is a matrix map, hence continuous (4.2.8). Since 1 has bounded sections in bv, $x^{(m)} = T1^{(m)}$ is bounded in X (8.0.1).

I do not know what conditions on X would correspond to $M(X)$ being sc, coregular etc.

18. THEOREM. *Let X be a BK space $\supset \phi$. If X has AB then X^f has AB. The converse is false in general but is true if $X \subset X^{ff}$. In particular the converse holds if X has AD.*

First, $1 \in B[M(X)]$ (Lemma 17) $\subset B[M(X^f)]$ since $M(X) \subset M(X^f)$ (10.1.7) and B is monotone (10.2.9). The result follows by Lemma 17. The falsity of the converse is shown in 7.2.5. The last sentence is by 7.2.15 — it is strictly a particular case by 7.2.16.

Finally, assume that X^f has AB and $X \subset X^{ff}$. Let $x \in X$. Then

$$\|x^{(m)}\| = \sup\{|\sum_{k=1}^{m} u_k x_k| : \|u\|_f \leq 1\} \text{ (10.1.6).}$$

Also,

$$|\sum_{k=1}^{m} u_k x_k| = |u^{(m)}x| \leq \|u^{(m)}\|_f \cdot \|x\|_{ff} \text{ (10.1.6).}$$

By Theorem 13, $\|u^{(m)}\| < M$ so $\{x^{(m)}\}$ is bounded.

A corollary is that *if X^f has AB, then X has AB iff $X \subset X^{ff}$*. Sufficiency is by Theorem 18; necessity is because $X^{ff} = X^{f\gamma} = X^{\gamma\gamma}$ by Theorems 8 and 18. Another is that *X^f has AB iff $\overline{\phi}$ has AB* by 7.2.4 and Theorem 18.

A useful smoothness property of AB spaces is that they contain $\{\delta^n\}$ nicely:

19. THEOREM. *Let X be an FK space such that $B \supset \overline{\phi}$. (See Corollary 21). Then $\overline{\phi}$ has AK and $S = W = \overline{\phi}$. This holds in particular if B is closed e.g. if X has AB. A special*

case is AD·AB implies AK.

Suppose first that X has AB. Define $f_n : X \to X$ by $f_n(x) = x - x^{(n)}$. Then $\{f_n\}$ is pointwise bounded, hence equicontinuous (7.0.2). Since $f_n \to 0$ on ϕ then also $f_n \to 0$ on $\overline{\phi}$ (7.0.3). This is the desired conclusion. If $B \supset \overline{\phi}$, apply the result just proved to the FK space $\overline{\phi}$ which has AB by Theorem 5. Since S is the set of x with AK it follows that $\overline{\phi} \subset S$. The opposite inclusion is 10.2.6.

20. LEMMA. *Let X be a BK space* $\supset \phi$ *and* $E \subset X$. *Then* $B \supset \overline{E}$ *(closure in X) iff there exists C such that*

$$\|x^{(m)}\| \leq C \cdot \|x\| \quad \text{for all } m \tag{1}$$

for all $x \in E$ *i.e.* $T_{B^+}|E \subset T_X|E$.

Necessity: $(\overline{E}, T_X)$ is an FK space $\subset B^+$. Hence for all $x \in \overline{E}$, $\|x\|_{B^+} \leq C\|x\|$ by 4.2.4. The result follows by Theorem 13.

Sufficiency: Let $x \in E$ with $\|x\| \leq 1$. Then $\|x\|_{B^+} \leq C$ by Theorem 13. By 8.2.3, $B^+ \supset \overline{E}$ so $B \supset \overline{E}$. (That $E \subset X \cap B^+$ is immediate from the hypothesis.)

21. COROLLARY. *Let X be a BK space* $\supset \phi$. *Then X has AB, B is closed in X,* $B \supset \overline{\phi}$ *iff (1) holds for all x, respectively, in X (or a dense subset), B,* ϕ.

22. EXAMPLE. Let X be a BK space with AD but not AK. (5.2.5, 5.2.14) this is *an AD space with* $X^\gamma \neq X^f$, *a fortiori* $X^\beta \neq X^f$, by Theorems 8 and 19. Also *B is not closed in X* (Theorem 19), *it is dense* since it includes ϕ. This example shows also that $X \subset X^{ff}$ *is not sufficient that X have AB.* (See the remarks following Theorem 18, and 7.2.15.)

The ideas of this section lead to a useful smoothing theorem:

23. THEOREM. *Let* X *be a* BK *space with* AD. *Then there exists a* BK *space* Y *with* AK *such that* $Y^\beta = X^\beta$.

Such Y is unique and $Y \supset X$ (8.6.1, 7.2.7). Let Y be the closure of ϕ in $X^{\beta\beta}$. Then Y has AK by Theorems 12, 19 and 10.1.5. Also since $Y \subset X^{\beta\beta}$ we have

$$Y^\beta \supset X^{\beta\beta\beta} \supset X^\beta \supset Y^\beta$$

using 7.2.2.

24. There exist non-trivial examples as in Example 22.

25. EXAMPLE. *A* BK *space with* AD *and* X *not closed in* $X^{\beta\beta}$*; a* BK *space with* AB *and the same property.* These are Examples 22 and 7. Clearly an AD space X is closed in $X^{\beta\beta}$ iff it has AK. For an AB space the condition is that it have a monotone norm.

26. To extend the results to FK spaces we may consider $s: \omega \to X^N$ given by $s(z) = \{z^{(n)}\}$. Here X^N is the space of sequences of members of X. Then $B^+ = s^{-1}[\ell^\infty(X)]$ with obvious meaning. Thus B^+, B are FK spaces. Another possibility is to apply techniques like those of Theorem 13 to each defining seminorm as was done in Theorem 1 and Example 2. Any form of Theorem 18 will have to depart from our present ideas, for example if $X = \omega$ then $X^f = \phi$ which is not an FK space (4.0.5).

27. REMARK. It should be emphasized that *the* BK *topologies on* B *and the other spaces are rarely used; mostly these spaces are treated as subspaces of* X *with the relative topology.* This is why the preceding Remark was made so casually. The first use of the topology of B^+ was in Theorem 13. This is unambiguous

since usually B^+ is not a subspace of X, thus inherits no topology from it. In Corollary 14 the B^+ norm is being used; as a subspace of X, B is not an FK space unless it is closed in X. These are the only places where the BK topologies of B^+, B are used.

References: [33], [59], [61].

10.4. THE SUBSPACE F.

1. DEFINITION. *An FK space X is said to have FAK (functional AK) if $F^+ \supset X$ i.e. $F = X$.*

2. THEOREM. *Let X be an FK space $\supset \phi$. Then $F^+ = X^{f\beta}$.*

This is proved in the same way as was 10.3.4, with cs instead of bs.

3. COROLLARY. *Let X be an FK space $\supset \phi$. Then F^+ is the same for all FK spaces Y between $\overline{\phi}$ and X i.e. $\overline{\phi} \subset Y \subset X$ implies $F^+(Y) = F^+(X)$. (The closure of ϕ is calculated in X).*

The proof is similar to that of 10.3.5.

4. COROLLARY. *Let X be an FK space $\supset \phi$. Then X has FAK iff $X^f \subset X^\beta$ i.e. $X^f = X^\beta$.*

The proof is similar to that of 10.3.8.

An easy corollary is that *for AD spaces, FAK and AB are equivalent* (7.2.7, 10.3.8).

Since $F \subset B$, FAK implies AB but not conversely:

5. EXAMPLE. *bv does not have FAK* for $F^+(bv) = F^+(bv_o) = bv_o^{f\beta} = bv_o^{\beta\beta} = bv_o$. Thus $F(bv) = bv_o$. (We used Corollary 3 and

the fact that bv_o has AK.) An instructive proof is: bv is not sc (9.2.6) thus $1 \notin F^+$ (10.2.7) so $1 \in bv \backslash F^+$. However *bv does have AB;* $B^+ = B = bv$ by a similar argument. Similarly $F^+(bs) = F(bs) = cs$, $B^+(bs) = bs$ so *bs has AB but not FAK.*

6. EXAMPLE. *X is sc iff $F^+ \supset bv$.* Since $bv = 1^{\beta\beta}$ and $F^+ = X^{f\beta}$ it is clear that $bv \subset F^+$ iff $1 \in F^+$. The result follows by 10.2.7.

7. EXAMPLE. *$X \supset c_o$ iff $F^+ \supset \ell^\infty$.* Necessity: $F^+(X) \supset F^+(c_o)$ by the monotonicity Theorem 10.2.9 and the latter is $c_o^{f\beta} = \ell^\infty$. Sufficiency: $\ell = (\ell^\infty)^\beta \supset (F^+)^\beta = X^{f\beta\beta} \supset X^f$ so $X \supset c_o$ by 8.6.2. Also *$F^+(c) = F^+(\ell^\infty) = F^+(c_o) = \ell^\infty$* by Corollary 3.

8. EXAMPLE. *Monotone norm does not imply FAK* (Example 5) *FAK does not imply X closed in $X^{\beta\beta}$* by 10.3.7; compare 10.3.25.

9. THEOREM. *Let X be an FK space $\supset \phi$. Then $B \supset \overline{F}$ (closure in X) iff F is closed in X. Thus B closed implies F closed.*

Sufficiency is trivial. Now suppose that $B \supset \overline{F}$. Fix $f \in X'$ and define $g: \overline{F} \to \ell^\infty$ by $g(x) = \{f(x^{(n)})\}$. Then $P_n \circ g: \overline{F} \to K$ given by $P_n[g(x)] = f(x^{(n)}) = \sum_{k=1}^{n} f(\delta^k)x_k$ is continuous, $\overline{F}$ having the relative topology of X. Thus $g: \overline{F} \to \ell^\infty$ is continuous (4.2.3), hence $g^{-1}[c]$ is closed in $\overline{F}$, and so, consequently, is $F = \cap\{g^{-1}[c]: f \in X'\}$.

A criterion for Theorem 9 is given in 10.3.20.

10. If X is a BK space Theorem 9 can be improved to: *F^+ is a closed subspace of B^+ and F is a closed subspace of B* in the norms given in 10.3.13. This is by 4.3.17, Theorem 2 and 10.3.4. Thus *F^+ is a BK space with monotone norm.* (For FK spaces see

10.3.26, and for a caveat, 10.3.27.) Theorem 9 then follows since $F \subset \overline{F} \subset B$ and so F is closed in $\overline{F}$ by 4.2.5.

An exact description of FAK spaces can be given; namely $X = \overline{\phi} \cup E$ where $\overline{\phi}$ has AK and E is made up of points selected from $(\overline{\phi})^{\beta\beta}$. (Theorem 12). A similar discussion could be given for AB.

11. LEMMA. *Let X be an FK space in which $\overline{\phi}$ has AK. Then $F^+ = (\overline{\phi})^{\beta\beta}$.*

For $F^+ = X^{f\beta}$ (Theorem 2) $= (\overline{\phi})^{f\beta}$ (7.2.4) $= (\overline{\phi})^{\beta\beta}$ (7.2.7).

12. THEOREM. *Let X be an FK space $\supset \phi$. Then X has FAK iff $\overline{\phi}$ has AK and $X \subset (\overline{\phi})^{\beta\beta}$.*

Necessity: X has AB since $F \subset B$ so $\overline{\phi}$ has AK by 10.3.19. The rest is by Lemma 11. Sufficiency is by Lemma 11.

Thus bv fails to have FAK because, although $\overline{\phi} = bv_o$ has AK, bv contains $1 \notin bv_o^{\beta\beta} = bv_o$. Similarly $\ell \oplus 1$, $c_o \oplus z$ $(z \notin \ell^\infty)$ fail to have FAK.

If $\overline{\phi}$ has AK we can see how much Corollary 4 fails by using the formula (whose easy proof we omit) $X^\beta = X^f \cap (X\backslash\overline{\phi})^\beta$.

13. If X has AK and is a β space (i.e. $X = X^{\beta\beta}$), then X is maximal FAK in the sense that X is not a proper closed subspace of any FAK space. This is immediate from Theorem 12 and applies to ℓ, bv_o, cs. However c_o is a closed subspace of any X such that $c_o \subset X \subset \ell^\infty$ (4.2.5) and each such X has FAK by Theorem 12 or because it has AB.

The conditions of Theorem 12 imply that $S = \overline{\phi}$. But making this substitution allows us to prove the seemingly stronger result that X has FAK iff $X \subset S^{\beta\beta}$ (Theorem 14). Certainly we cannot omit the condition $\overline{\phi}$ has AK from Theorem 12, witness any AD space without AK. (5.2.5 or 5.2.14.)

14. THEOREM. *Let X be an FK space $\supset \phi$. These are equivalent: (i) X has FAK, (ii) $X \subset S^{\beta\beta}$, (iii) $X \subset W^{\beta\beta}$, (iv) $X \subset F^{\beta\beta}$, (v) $X^{\beta} = S^{\beta}$, (vi) $X^{\beta} = F^{\beta}$.*

That (ii) implies (iii) and (iii) implies (iv) are trivial since $S \subset W \subset F$. If (iv) is true, then $X^{\beta} \supset F^{\beta} = X^{f\beta\beta} \supset X^{f}$ so (i) is true by Corollary 4. If (i) is true, Theorem 12 implies that $S = \overline{\phi}$ and that (ii) is true. The equivalence of (v), (vi), (vii) with the others is clear.

References: [33], [59], [61].

10.5. THE SUBSPACE W.

Recall that $W = \{x\colon f(x) = \sum f(\delta^k)x_k$ for all $f \in X'\}$. Thus $W \subset X$. Members of W are said to have SAK, and if $W = X$, X is said to have SAK. S is for Schwach = Weak. It turns out that the concept of SAK space is redundant:

1. THEOREM. *Let X be an FK space $\supset \phi$. These are equivalent: (i) X has SAK, (ii) X has AK, (iii) $X^{\beta} = X'$.*

The meaning of condition (iii) will be explained during the proof. Clearly (ii) implies (i). Conversely if X has SAK it must have AD for $W \subset \overline{\phi}$ by 10.2.6. It also has AB since $W \subset B$. Thus X has AK by 10.3.19. That (ii) implies (iii) is 7.2.9 where also the meaning of (iii) is given. Conversely if (iii) holds, let $f \in X'$; there exists $u \in X^{\beta}$ such that $f(x) = ux$ for $x \in X$. Since $f(\delta^n) = u_n$ it follows that each $x \in W$ i.e. X has SAK.

The discussion of c in 7.2.11 is pertinent here.

Very sharp conditions can be given for W to be closed in X:

2. THEOREM. *Let* X *be an* FK *space* $\supset \phi$. *These are equivalent: (i)* W *is closed in* X, *(ii)* $B \supset \overline{\phi}$, *(iii).* $F \supset \overline{\phi}$, *(iv)* $W = \overline{\phi}$, *(v)* $S = \overline{\phi}$, *(vi)* S *is closed in* X. *If* B *is closed in* X *all these hold.*

(ii) implies (v): By 10.3.19, $\overline{\phi}$ has AK, i.e. $\overline{\phi} \subset S$. The opposite inclusion is 10.2.6. That (v) implies (iv), (iv) implies (iii) and (iii) implies (ii) are because $S \subset W \subset \overline{\phi}$, $W \subset F \subset B$; (i) implies (iv) and (vi) implies (v) since $\phi \subset S \subset W \subset \overline{\phi}$. Finally (iv) implies (i) and (v) implies (vi).

Example 13.2.24 shows that $W = S$ is not sufficient here even for c_A with A conservative.

Condition (ii) was discussed in 10.3.21.

3. If X is a BK space parts of this result are improved by: *W is a closed subspace of F and B* with the norms given in 10.3.13. since $W = \cap h^{\perp}$ where, for any $f \in X'$, $h = h_f$ is defined on F by $h(x) = f(x) - \sum f(\delta^k)x_k$, and is continuous by the Banach-Steinhaus theorem (1.0.4). (See 10.4.10.) Hence *W is a BK space with AB.* If *W has AD* it must have AK by 10.3.19, hence has AK with the (smaller) topology of X and so *$W = S$.* We shall see that X may have AD and W not. (12.5.6).

Sources: [33], [61], [86], [87].

10.6. BASIS

There is a fairly difficult Functional Analysis theorem which says that a weak basis for a ℓ.c. Fréchet space is a Schauder basis i.e. if $x = \sum t_k b^k$ uniquely, the series converging weakly, then the equality holds in the Fréchet topology, and the maps $x \to t_k$ are continuous. See [80] #5.4.302. The difficult part of the proof is the second assertion. Since we are dealing with FK spaces, continuity of coordinates is assumed and the whole theorem becomes easy. (A similar point was made in 7.1.5):

1. THEOREM. *Let* X *be an* FK *space* $\supset \phi$. *Suppose that for each* x, $\sum x_k \delta^k$ *converges weakly to* X. *Then* X *has* AK.

This is 10.5.1.

2. THEOREM. *Let* X *be an* AD *space and suppose that the series* $\sum x_k \delta^k$ *is bounded for each* $x \in X$. *Then* x *has* AK.

This is 10.3.19 since the hypothesis is that each x has bounded sections.

For instance, in Theorem 2 it would be sufficient to assume that the series is weakly Cauchy. (8.0.2). This just gives the special case AD. FAK implies AK.

3. COROLLARY. *Let* X *be an* FK *space* $\supset \phi$. *If* $X^\gamma = X^f$ *(in particular if* $X^\beta = X^f$*) then* $\overline{\phi}$ *has* AK.

This is 10.3.19 since X has AB (10.3.8).

4. COROLLARY. *Let* X *be an* AD *space. Then* X *has* AK *iff* $X^\gamma = X^f$.

CHAPTER 11
EXTENSION

11.0. FUNCTIONAL ANALYSIS

1. Let E be a dense subspace of a metrizable vector space X, Y a Fréchet space and $T: E \to Y$ a continuous linear map. Then T has an extension to a continuous linear map: $X \to Y$. [80], #2-1-11.

11.1. ROW FUNCTIONS

Extension of functions is an important topic. The Hahn-Banach theorem (3.0.1), which extends functionals, does not apply to maps of Banach spaces, for example the identity map: $c_o \to c_o$ cannot be extended to ℓ^∞. ([80], Example 14.4.9). The problem of which spaces allow extensions is a well-studied one which will not concern us. The best result which holds in all situations is: if $T: E \to Y$, $E \subset X$, is continuous and linear, then T has an extension to $T: \overline{E} \to Y$. (11.0.1.) The problem we shall study is to inquire when such an extension of a matrix is a matrix. This is really a question about the rows of A so we begin by isolating this aspect.

1. DEFINITION. *A functional* h *on a vector space* $E \supset \phi$ *is called a row function if* $h = \hat{u}$ *on* E, *i.e.* $h(x) = ux$, *for some* $u \in E^\beta$. *Equivalently,* $h(x) = \sum h(\delta^k)x_k$ *for* $x \in E$.

Once a row function $h = \hat{u}$ has been defined on ϕ it immediately has a natural domain, namely u^β; i.e. $h(x) = ux$ is defined for such x, and h has a unique extension *as a row function* to this domain. Thus several questions arise if h is a row function defined on ϕ and X is an FK space $\supset \phi$. First: how is the natural domain of h located relative to X? Second: on how much of the domain is h continuous? Third: if h is continuous on ϕ what is its Hahn-Banach extension to all of X?

We begin with a special case of continuity. This result justifies taking $u \in X^f$.

2. THEOREM. *Let X be an FK space $\supset \phi$ and $h = \hat{u}$ a row function on ϕ. Then h is X-continuous on ϕ (i.e. $h|\phi$ is continuous in the relative topology of X) iff $u \in X^f$.*

Necessity: Extend h to $f \in X'$ by the Hahn-Banach theorem (3.0.1). Then $u_n = f(\delta^n)$. Sufficiency: Say $u_n = f(\delta^n)$. Then for $x \in \phi$, $h(x) = ux = f(x)$ so $h = f$ on ϕ i.e. h is continuous.

Of course $h \in X'$ iff $u \in X^\beta$. Sufficiency is by the Banach-Steinhaus theorem (1.0.4).

The situation with respect to continuity is bizarre, namely a row function can be continuous on a proper subset of its natural domain.

3. THEOREM. *Let X be an FK space $\supset \phi$ and h an X-continuous row function on ϕ (so $h = \hat{u}$ with $u \in X^f$). Then the natural domain of h includes F^+ but h need not be continuous on F. However h must be continuous on W. The natural domain of h need not include $\overline{\phi}$.*

Viewed as an extension theorem this says that h has a continuous extension as a row function to W and an extension as a row function to F, not necessarily continuous. It need not have any extension as a row function to $\overline{\phi}$. Of course h has a continuous extension to all of X by the Hahn-Banach theorem or, more directly, since $h = f|\phi$ for some f, by hypothesis.

First the natural domain of h is $u^\beta \supset X^{f\beta} = F^+$ (10.4.2) since $u \in X^f$ by Theorem 2. Second h is continuous on W for let $f \in X'$ with $f(\delta^n) = u_n$. For $x \in W$, $f(x) = \sum f(\delta^k)x_k = ux = h(x)$ i.e. $h = f$ on W.

The rest of Theorem 3 will be shown in examples. In both cases X may be chosen to be c_A with A a conservative coregular triangle.

4. EXAMPLE. *A continuous row function with no extension to* $\overline{\phi}$ *(as a row function.)* Let X be an AD space and $u \in X^f \backslash X^\beta$ (10.3.22). Say $u_n = f(\delta^n)$, $f \in X'$. Then f is not a row function since $u \notin X^\beta$, but $f|\phi$ is a row function, as are all linear functionals on ϕ.

5. EXAMPLE. *A row function continuous on* W *but not on* F. Let X be an AD space such that $W \neq F$. Any coregular space has the latter property by 10.2.7; to make it AD is more difficult, an example is the convergence domain of the conservative coregular non-replaceable triangle given in 5.2.5. Let $x \in F\backslash W$. There exists $f \in X'$ such that $f(x) \neq ux$ where $u_n = f(\delta^n)$, by definition of W. (In the case of 5.2.5 we may take $x = 1$, $f = \lim_A$, $u = a$. See 5.2.1 (v).) Since $u \in X^f$, $\hat{u}$ is continuous on ϕ (Theorem 2), as is f. They are equal on ϕ but not at x. Since X has AD this shows that $\hat{u}$ is not continuous.

If $u^\beta \cap X$ is closed in X, then $\hat{u}$ is continuous on its domain by the Banach-Steinhaus theorem 1.0.4 (so $u \in X^f$). The converse is false by the next two examples. These examples show an extension of $\hat{u}$ to all of X of a very special form.

6. EXAMPLE. Let $(Ax)_n = x_n + x_{n-1}$ so that $\frac{1}{2}A$ is (except for one row) the regular triangle Q (1.2.5), $X = c_A$. *Let* $u \in X^f$. *Then* $\sum u_k x_k$ *is (C,1) summable for all* $x \in X$. Let $y_k = \sum_{i=1}^{k} u_i x_i$. Now if $u_k = f(\delta^k)$ with $f(x) = \mu \lim_A x + t(Ax)$ (4.4.5) it follows that $u_i = t_i + t_{i+1}$ and so, by Abel's identity (1.2.9),

$$y_k = \sum_{i=1}^{k} t_i(x_i + x_{i-1}) + t_{k+1}y_k = v_k + w_k,$$

say. Now $v_k \to t(Ax)$ hence *v is (C,1) summable to t(Ax)* since (C,1) is regular, and we shall show that *w is (C,1) summable to 0.* To see this, observe that $(1/n)\sum_{k=1}^{n} w_k = (Et)_n$ where $e_{nk} = x_k/n$ for $k \le n$, $t = (t_2, t_3, t_4, \ldots)$. The matrix E has null columns and $\sup|e_{nk}| = \sup\{|x_k/n| : k \le n\} < \infty$ since $x_n = o(n)$ by 3.3.11. Thus $E \in (\ell : c_0)$ by 8.4.1A.

7. EXAMPLE. *In Example 6,* $ux = t(Ax)$ *for all* $x \in u^\beta \cap X$. For it was proved that the (C,1) sum of the series is t(Ax) for all x, hence if the series is convergent, this is its sum since (C,1) is regular. This shows that $\hat{u}$ *is continuous on its domain* since t(Ax) is continuous in x on X. Finally *this domain need not be closed* for $u^\beta \supset c$ since $u \in \ell$, c is dense since A is of type M hence perfect (3.3.4), and *the domain is a proper subset of X* if u is chosen in $X^f \setminus X^\beta$ (See the next Remark): let $x_k = (-1)^k \log k$ and choose $t \in \ell$ so that $\lim t_{k+1}x_k$ does not exist; then y (Example 6) is divergent. With $u_k = t_k + t_{k+1}$, $f(x) = t(Ax)$, $f(\delta^k) = u_k$ and so u has the required property.

8. The existence of u in the latter part of Example 7 can be deduced from general theorems: X does not have FAK by

13.4.10, 13.3.1, so $X^f \neq X^\beta$ by 10.4.4. A very similar situation may be observed in 5.2.14 where a construction much like that just given could be avoided by an appeal to a general theorem.

9. A row function composed with a matrix need not be a row function (1.4.6).

11.2. MATRICES

Suppose that X, Y are FK spaces with $X \supset \phi$ and that A is a matrix mapping ϕ into Y i.e. the columns of A belong to Y. That A be X-continuous it is necessary that each row of A belong to X^f since $P_n \circ A$ is continuous, using 11.1.2. It is not sufficient, however, e.g. take $X = c_o$, $Y = \ell$, $A = I$. Thus the continuity of A is a stronger assumption.

Such A can be extended to $T: \overline{\phi} \to Y$ (11.0.1), and the question arises whether T is given by a matrix (which must be A since $T = A$ on ϕ.) This is equivalent to asking whether $A[\overline{\phi}] \subset Y$. Similarly the conditions $A[W] \subset Y$, $A[F] \subset Y$ are equivalent to $T = A$ on W, F i.e. A extends to W, F as a matrix.

In this section the subspaces $\overline{\phi}$, F, W are subspaces of X.

1. THEOREM. *Let X be an FK space $\supset \phi$, Y an FK space and A a matrix such that $A: \phi \to Y$ is X-continuous. Then A extends to W as a matrix, (equivalent statements are $A[W] \subset Y$, $Y_A \supset W$), and A is X-continuous on W. Also $\omega_A \supset F$ but $A[F]$ need not be a subset of Y even if W is dense in F (thus A need not be X-continuous on F). Finally ω_A need not include $\overline{\phi}$ (in particular $A[\overline{\phi}] \not\subset Y$).*

Extend A to $T: \overline{\phi} \to Y$ (11.0.1). Applying 11.1.3 to the nth row of A and $P_n \circ T$ for each n shows that $T = A$ on W, hence $A[W] \subset Y$, and $\omega_A \supset F$ since each row of $A \in F^\beta$ by 11.1.3. The rest of Theorem 1 will be shown in examples. In each case X may be chosen to be c_M with M a conservative coregular triangle, $Y = c_o$.

2. EXAMPLE. *$\omega_A \supset \overline{\phi}$ in Theorem 1.* This is the same as 8.3.5.

3. EXAMPLE *$A[F] \not\subset Y$ in Theorem 1.* Consider first the simple example of the identity matrix which maps ϕ into c_o but does not map c into c_o. Here $W = c_o$, $F = c$. To give an example with W dense in F requires more effort. Let $X = c_M$ where M is a conservative coregular non-replaceable triangle as given in 5.2.5. Let $m_k = \lim m_{nk}$, $m = \{m_k\}$. Let $a_{nk} = m_{nk} - m_k$, $Y = c_o$. For $x \in \phi$, $(Ax)_n = (Mx)_n - mx$ so that $Ax = Mx - (mx)1$. Now $M: X \to c$ is continuous, also $\hat{m}$ is X-continuous on ϕ by 11.1.2 since $m_k = \lim_M \delta^k$ and $\lim_M \in X'$. Thus $A: \phi \to c$ is X-continuous, hence so is $A: \phi \to c_o$. But $(A1)_n = (M1)_N - \sum m_k \to \chi(M) \neq 0$ so $A1 \notin c_o$ but $1 \in F$ (10.2.7). Also X has AD so W is dense.

With more assumptions the extension can be made:

4. THEOREM. *In Theorem 1 assume also that $\omega_A \supset \overline{\phi}$. Then A extends to $\overline{\phi}$ as a matrix i.e. $A[\overline{\phi}] \subset Y$, and $A: \overline{\phi} \to Y$ is X-continuous. In particular this holds if A is row finite.*

Let $x \in \overline{\phi}$; say $x^m \to x$, $x^m \in \phi$. Then $\{Ax^m\}$ is Cauchy, hence convergent, in Y. Let its limit be called y. Now $x^m \to x$ in ω_A. (Apply 4.2.4 with X replaced by $\overline{\phi}$.) Hence $Ax^m \to Ax$ in ω since $A: \omega_A \to \omega$ is continuous (4.2.8). Also $Ax^m \to y$ in ω since Y is an FK space. It follows that $Ax = y \in Y$. The continuity of A follows from 4.2.8.

It does not follow that $A[F] \subset Y$ even if A is row finite, as is shown by the first part of Example 3.

Lemma 8.2.1 is the special case of Theorem 4 in which A is the identity matrix.

One might try to improve Theorem 4 by concluding that in Theorem 1 the extension can be made to $\omega_A \cap \overline{\phi}$. That this is false is shown by Example 3 since this subspace $\supset F$ there.

With the notation of Example 3 define $g_{nk} = m_k$ for $k \leq n$, 0 for $k > n$. Then G is a triangular matrix. It follows from Theorem 4 that $G: \phi \to c$ is not X-continuous since $(M-G)1 \notin c_0$.

There is something paradoxical about Example 2; namely $A: \phi \to Y$ is X-continuous. If Y_A induces a smaller topology on ϕ than that of X, it follows that $Y_A \supset \overline{\phi}$. (8.2.1). The fact is that the continuity of A does not imply this. More on this in the following optional remark.

5. REMARK (This may be omitted). The *weak topology by* A on Y_A, written w_A, is defined to be the smallest topology such that $A: Y_A \to Y$ is continuous [80], Example 4-1-9. For $A = 0$, w_A is the indiscrete topology — thus far from the FK topology of $Y_A = \omega$. The map $A: \phi \to Y$ is X-continuous iff $w_A \subset T_X$ on ϕ, whereas $Y_A \supset \overline{\phi}$ iff the FK topology of $Y_A \subset T_X$ on ϕ. (8.2.1). If A is row-finite, w_A and T, the FK topology of Y_A, share this property: *let* $E \subset \phi$ *and suppose that* E *is bounded in* ω. *Then* E *is* w_A *bounded* (i.e. $A[E]$ is bounded in Y) *iff* E *is bounded in* Y_A. Since $w_A \subset T$, half of this is trivial. Conversely, if E is w_A bounded it is T bounded by 8.3.1. This can be used to prove the row-finite case in Theorem 4; if E is bounded in X it is bounded in ω and $A[E]$ is bounded in

Y. Thus by 8.3.1, E is bounded in Y_A and the result follows by 8.2.3.

Source: [86].

CHAPTER 12
DISTINGUISHED SUBSPACES OF MATRIX DOMAINS

12.0. FUNCTIONAL ANALYSIS

1. If a Banach space has the property that weakly convergent sequences are norm convergent the same is true for Cauchy sequences. [80], #8-1-10.

2. If a Banach space Y has a closed maximal subspace with the property given in 1; then Y also has this property. Proof: Say $Y = X \oplus u$. Let $y_n = x_n + t_n u \to 0$ weakly. Let $f \in Y'$, $f = 0$ on X, $f(u) = 1$ (5.0.1). Then $t_n = f(y_n) \to 0$. For $g \in X'$ extend g to all of Y (Hahn-Banach, 3.0.1). Then $g(x_n) = g(y_n) - t_n g(u) \to 0$ so $x_n \to 0$ in norm. Hence also does y_n.

12.1. MATRIX DOMAINS.

The original ground space of summability is c_A. More general FK spaces were covered in Chapter 10; in this chapter we discuss Y_A. Its properties depend on the choice of Y and the choice of A; our procedure will be to fix Y and discuss how the properties of Y_A depend on those of A. This discussion will depend on which Y is chosen. In each of the subsequent sections of this chapter Y will be specialized in some way. The historic case $Y = c$ will be taken up in §5 and Chapter 13.

We pause to note that matrix maps need not preserve distinguished subspaces. For example, let X be any FK space $\supset \phi$, $Y = \ell \oplus 1$, $a_{n1} = 1$, $a_{nk} = 0$ for $k > 1$. Then $A: X \to Y$, $\delta^1 \in B(x)$ but $A\delta^1 = 1 \notin B(Y)$ since $\|1^{(n)}\| = n$. (In particular X can be $Y_A = \omega$ here.)

To characterize distinguished subspaces of Y_A is quite trivial, for example $z \in B^+$ iff $\{p(z^{(m)})\}$ is bounded for each seminorm defining the topology of Y_A. These may be read from 4.3.12. It turns out however that a simpler criterion is available, namely that $\{Az^{(m)}\}$ is bounded in Y. (Theorem 3). This is simpler because Y has fewer seminorms than Y_A (4.3.12) and fewer functions in its dual (4.4.2); and more useful because properties of Y will force similar properties on Y_A. A specific instance of this is shown in 12.4.3, 12.4.7. The basic reason for the simplification is that the seminorms of Y_A are made up of those of ω_A and those inherited from Y. Since ω_A has AK (4.3.8), it offers no obstacle to boundedness or convergence, leaving the issue to be decided by the seminorms of Y. This proof is used in Theorem 6 but in Theorems 3, 4, 5 other techniques are more natural at this stage.

1. REMARK. In this section, *$z \in \omega$, Y is an FK space, and A is a matrix such that $Y_A \supset \phi$* i.e. the columns of A belong to Y. The subspaces *S, W, F, B are calculated in the FK space Y_A.*

2. LEMMA. *With the notation of Remark 1,* $Az^{(m)} = \sum_{k=1}^{m} z_k a^k$ *where a^k is the kth column of A.*

For $(Az^{(m)})_n = \sum_{k=1}^{m} a_{nk}z_k = (\sum z_k a^k)_n$.

3. THEOREM. *With z, Y, A as in Remark 1, these are equivalent:*

(i) $z \in B^+$,

(ii) $\{Az^{(m)}\}$ *is bounded in* Y,

(iii) $Y_{A\cdot z} \supset bv_0$,

(iv) $\{z_k g(a^k)\} \in bs$ *for each* $g \in Y'$, *where* a^k *is the* kth

column of A.

Also these are equivalent: $z \in B$, $Y_{A\cdot z} \supset bv$, *(ii) and* $z \in Y_A$, *(iv) and* $z \in Y_A$.

(i) = (iii): by 10.2.7(i) since $z^{-1}\cdot Y_A = Y_{A\cdot z}$. (iii) = (ii) by the last part of 8.6.4 since the kth column of $A\cdot z$ is $z_k a^k$. (ii) = (iv): (ii) is true iff $\{g(Az^{(m)})\}$ is bounded for each $g \in Y'$. (8.0.2). By Lemma 2, this is the same as (iv).

The second set of equivalences is clear since $z \in Y_A$ iff $1 \in Y_{A\cdot z}$.

4. It is possible to make these results look like those of §10.3. Suppose, with A fixed, we define Y^g to be the set of all sequences $\{g(a^k)\}$ for $g \in Y'$. Then condition (iv) says $B^+ = Y^{g\gamma}$, very like 10.3.4. The analogue of 10.3.8 would be: Y_A has AB if $Y^g \subset Y_A^\gamma$. Similar remarks hold for F in Theorem 5.

5. THEOREM. *With* z, Y, A *as in Remark 1 these are equivalent:*

(i) $z \in F^+$,

(ii) $\{Az^{(m)}\}$ *is weakly Cauchy in* Y *i.e.* $\{g[Az^{(m)}]\} \in c$ *for each* $g \in Y'$,

(iii) $Y_{A\cdot z}$ *is sc,*

(iv) $\{z_k g(a^k)\} \in cs$ *for each* $g \in Y'$ *where* a^k *is the kth column of* A.

Also these are equivalent: $z \in F$, $Y_{A\cdot z}$ *is vsc, (ii) and* $z \in Y_A$, *(iv) and* $z \in Y_A$.

(i) = (iii) by 10.2.7 (iii); (iii) = (ii) by 9.4.1 since the kth column of $A\cdot z$ is $z_k a^k$; (ii) = (iv) by Lemma 2. The last part is clear as in Theorem 3.

A weak form of part of the next result was given in 5.6.8.

6. THEOREM. *With* z, Y, A *as in Remark 1 these are equivalent:*

(i) $z \in W$,

(ii) $Az^{(m)} \to Az$ *weakly in* Y,

(iii) $Y_{A \cdot z}$ *is conull,*

(iv) $\sum z_k g(a^k) = g(Az)$ *for each* $g \in Y'$.

(i) ≡ (iii) by 10.2.7 (v); (iii) = (ii) by 9.4.9 since the kth column of $A \cdot z$ is $z_k a^k$; (ii) = (iv) by Lemma 2.

7. THEOREM. *With* z, Y, A *as in Remark 1 these are equivalent:*

(i) $z \in S$,

(ii) $Az^{(m)} \to Az$ *in* Y,

(iii) $Y_{A \cdot z}$ *is strongly conull,*

(iv) $\sum z_k a^k = Ax$, *convergence in* Y, *where* a^k *is the kth column of* A.

(i) = (iii) by 10.2.7 (vi). (ii) = (iv) by Lemma 2. (i) implies (ii): $z = \sum z_k \delta^k$ and $A: Y_A \to Y$ is continuous so $Az = \sum z_k A\delta^k = \sum z_k a^k$. (ii) implies (i): First ω_A has AK by 4.3.8 hence $u(z-z^{(m)}) \to 0$ for any $z \in \omega_A$ where $u = p$ or h in 4.3.8 and 4.3.12. Thus $z \in S$ if $q[A(z-z^{(m)})] \to 0$ where q is a typical seminorm of Y by 4.3.12. But this is simply $Az^{(m)} \to Az$ in Y.

8. A host of invariance theorems can be read from these results. The subspaces S, W, F, B are invariant i.e. the same for matrices A, M if $Y_A = Z_M$; indeed they are named i.e. defined in terms of the FK space. Thus the other properties are invariant; let us give just one example. Suppose $Y_A = Z_M$, $z \in \omega$ and $Az^{(m)} \to Az$ weakly in Y; then $Mz^{(m)} \to Mz$ weakly in Z also. This follows from Theorem 6 (ii).

12.2. ASSOCIATIVITY

There are two subspaces of Y_A which play an important role in summability.

1. REMARK. In this section *Y is an FK space and A is a matrix such that $Y_A \supset \phi$. The distinguished subspaces are calculated in Y_A.*

2. DEFINITION. *With the notation of Remark 1, $L_e^+ = \{z \in \omega$: $(tA)z$ exists for all $t \in Y^\beta\}$, $L_e = L_e^+ \cap Y_A$,*

$$L_a = \{x \in Y_A : (tA)x = t(Ax) \text{ for all } t \in Y^\beta\}.$$

After this chapter Y will always be c.

The letter L was used because in summability one takes $Y = c$, $Y^\beta = \ell$. The e stands for existence, a for associativity. Note that $t(Ax)$ always exists since $Ax \in Y$, $t \in Y^\beta$; also $\phi \subset L_a \subset L_e$. Another fact which may be of interest some day is that the assumption $Y_A \supset \phi$ is not always needed.

We have seen these formulas before: 9.6.7 says that if $Y_A \supset bv$ and Y has AD, then $L_a \supset bv$, this was a key step in the proof of 9.6.8.; 1.4.4 shows that if $Y = \omega$, $L_a = \omega_A$ and if $A \in \Phi$ and $Y^\beta = \ell$ (e.g. $Y = c_0, c, \ell^\infty$) then $L_a \supset Y_A \cap \ell^\infty$, the latter played the same role for 4.4.7. Also 4.4.9 is an associativity result and was applied to prove 4.6.7.

3. $L_e^+ \subset \omega_A$ since taking $t = \delta^n$ yields $(tA)z = (Az)_n$. Theorem 4 implies then that $F^+ \subset \omega_A$. This also follows from the definition of F^+ using $f(x) = (Ax)_n$.

The key motivation is that if $t \in Y^\beta$ and $f \in Y_A'$ is defined by $f(x) = t(Ax)$ then $\sum f(\delta^k)x_k = (tA)x$ and this will exist for some x and be $f(x)$ for some x. In the latter case it gives good information, for example if $f = 0$ on ϕ then $f(x) = 0$ for such x.

4. THEOREM. *With the notation of Remark 1,* $F^+ \subset L_e^+$, $F \subset L_e$, $W \subset L_a$.

If $z \in F^+$, $t \in Y^\beta$ let $f(x) = t(Ax)$ define $f \in Y_A'$. By definition of F^+, $\sum f(\delta^k)z_k$ converges and this is $(tA)z$. If $x \in W$ the equality $f(x) = \sum f(\delta^k)x_k$ with the same f says $t(Ax) = t(Ax)$.

In case Y^β has AD this result can be sharpened. Note that this includes the all-important case $Y = c$. The "improvement of mapping" plays a significant role.

5. THEOREM. *Let* A *be a matrix and* Y *a* BK *space such that* Y^β *has* AD. *Then* $L_e = L_a$, $B^+ \cap \omega_A \subset L_e^+$, $B \subset L_e = L_a$.

That $L_e = L_a$ is simply 9.6.5 with $Z = Y^\beta$, $x \in Y_A$. Condition (ii) is automatic as pointed out just after Definition 2. Next let $z \in B^+ \cap \omega_A$. By 12.1.3, $A \cdot z \in (bv_0 : Y)$ and so by 8.3.8, $(A\cdot z)^T \in (Y^\beta : bs)$. Since $z \in \omega_A$, each row of $A \cdot z$ i.e. each column of $(A\cdot z)^T$ belongs to cs. Hence $(A\cdot z)^T \in (Y^\beta : cs)$ by the second part of 8.3.8 which improves the map. So, for $t \in Y^\beta$, $(A\cdot z)^T t \in cs$ yielding the convergence of $\sum [(A\cdot z)^T t]_k = \sum_k \sum_n a_{nk} z_k t_n = (tA)z$. Thus $z \in L_e^+$.

We can deduce 9.6.7 thus: $L_a = L_e = B(Y_A) \supset B(bv)$ (10.2.9) $= bv$.

6. Take $A = I$. Then $Y_A = Y$. Clearly $L_e = L_a = Y$, $L_e^+ = Y^{\beta\beta}$. By choosing Y suitably one sees that, (pace Theorems 4, 5), F, W, B are fairly independent of the L spaces. The case $A = I$ shows that *subsequent results of this chapter proved for* Y_A *will also hold for* Y *itself.*

7. EXAMPLE. *In Theorem 5,* $B^+ \not\subset L_e^+$, $B^+ \not\subset \omega_A$. Let $z \in bs\backslash cs$. Let $a_{1k} = 1$, $a_{nk} = 0$ for $n > 1$. Then $z \in B^+\backslash \omega_A$ so $z \notin L_e^+$ by Remark 3. Here Y can be any BK space; use 12.1.3, $Az^{(m)} = (\sum_{k=1}^m z_k)\delta^1$.

Recall that since $A \in (Y_A:Y)$ it follows that $A^T \in (Y^\beta:Y_A^f)$ (8.3.8). The next result shows when a sharper inclusion is true. (Not always — 13.2.1, 13.4.10).

8. THEOREM. *With the notation of Remark 1,* $A^T \in (Y^\beta:Y_A^\beta)$ *iff* $L_e = Y_A$.

Clearly $L_e = Y_A$ iff $tA \in Y_A^\beta$ for all $t \in Y^\beta$ and this says precisely that $A^T \in (Y^\beta:Y_A^\beta)$.

An important aspect of associativity is its role in classifying various members of the dual space.

9. DEFINITION. $G = \{f \in Y_A' : f(x) = t(Ax)$ *with* $t \in Y^\beta\}$.

Recall also the embedding: $X^\beta \to X'$ in which $u \to \hat{u}$ with $\hat{u}(x) = ux$. (7.2.9).

10. THEOREM. *With the notation of Remark 1,* $G \subset Y_A^\beta$ *iff* $L_a = Y_A$.

Necessity: Let $t \in Y^\beta$ and define $f \in G$ by $f(x) = t(Ax)$. By hypothesis $t(Ax) = \alpha x$ for all x. With $x = \delta^k$ this yields $(tA)_k = \alpha_k$ and so $t(Ax) = (tA)x$ for all x. Sufficiency: let $f \in G$, say $f(x) = t(Ax)$. By hypothesis this is $(tA)x$ so $x \in Y_A^\beta$.

12.3. AK SPACES

This section and the next one are of secondary importance.

1. REMARK. In this section *Y is an AK space and A is a matrix whose columns are in Y i.e.* $Y_A \supset \phi$. *The subspaces S, W, F, B are calculated in* Y_A.

2. THEOREM. *With* Y, A *as in Remark 1,* $L_e^+ = F^+$, $L_e = F$, $L_a = W$.

Let $z \in L_e^+$, $g \in Y'$. With $t_n = g(\delta^n)$ we have $t \in Y^\beta$ since Y has AK (7.2.7). Applying 12.1.2 and the fact that Y has AK,
$g(Az^{(m)}) = \sum_{k=1}^{m} z_k g(a^k) = \sum z_k \sum_n g(\delta^n)a_{nk} = \sum_{k=1}^{m} \sum_n z_k a_{nk} t_n \to (tA)z.$
Thus $z \in F^+$ by 12.1.5 (ii). If $z \in L_a$ the last expression is equal to $t(Az) = g(Az)$ so $z \in W$ by 12.1.6 (ii). The opposite inclusions were given in 12.2.4.

3. It follows that the three L subspaces are invariant as spelled out in 12.1.8 provided that Y, Z are AK spaces.

4. EXAMPLE. Let $Y = cs$, $(Ax)_n = x_n - x_{n-1}$. Then $Y_A = c$ so $L_e^+ = F^+ \equiv \ell^\infty$, $L_e = F = c$, $L_a = W = c_0$.

Sources: [13], [86], [87].

12.4. WEAK AND STRONG CONVERGENCE

This section is of secondary importance.

1. REMARK. In this section *Y is an FK space such that weakly convergent sequences are convergent in the FK topology, A is a matrix such that $Y_A \supset \phi$. The subspaces S, W, F, B are calculated in Y_A.*

2. EXAMPLE. ℓ has the property mentioned in Remark 1 (1.7.20). So also do bv_0 and bv because bv_0 is equivalent with ℓ and $bv = bv_0 \oplus 1$ (7.3.4, 12.0.2).

3. If Y is as in Remark 1, then for Y itself, $S = W = F = F^+$ since if $x \in F^+$, $\{x^{(n)}\}$ is weakly Cauchy, hence Cauchy in the FK topology of Y (12.0.1), so convergent, say $x^{(n)} \to y$.

since $x^{(n)} \to x$ in ω it follows that $y = x$, so $x \in S$.

If Y is also AK we can conclude that $Y^{\beta\beta} = Y$ since $Y^{\beta\beta} = Y^{f\beta}$ (7.2.7) $= F^+ = F \subset Y$. Thus c_o does not have the property of Remark 1.

4. THEOREM. *With Y, A as in Remark 1 suppose that A is row finite. Then Y_A has the property that weakly convergent sequences are convergent.*

Let $x^n \to 0$ weakly. Then $Ax^n \to 0$ weakly in Y (4.0.11) hence $Ax^n \to 0$ in Y. Consulting the seminorms of Y_A given in 4.3.12 we see that each h_n is redundant, $q \circ A(x^n) \to 0$ as just proved, and $p_k(x^n) = |x_k^n| \to 0$ for each k since Y_A is an FK space.

5. EXAMPLE. Let z be a sequence. Then $z^\alpha = \ell_A$ where $Ax = z \cdot x$ (4.3.17). Hence this space also has the property of Remark 1.

The result of Remark 3 holds for Y_A if A is row finite, by Theorem 4.

6. EXAMPLE. *Row finite cannot be omitted in Theorem 4.* Let $a_{1k} = 1$, $a_{nk} = 0$ for $n > 1$. Then $\ell_A = cs$. Let $u^n = \delta^n - \delta^{n+1}$. Then $\|u^n\| = 1$, yet $u^n \to 0$ weakly, for if $f \in cs'$, $\{f(\delta^n)\} \in cs^f = bv$ and $f(u^n) = f(\delta^n) - f(\delta^{n+1}) \to 0$ since $bv \subset c$ (7.3.2).

In spite of Example 6 the result of Remark 3 does hold for Y_A:

7. THEOREM. *With Y, A as in Remark 1, $S = W = F = F^+$.*

If $z \in F^+$, $(Ax^{(m)})$ is weakly Cauchy in Y by 12.1.5, hence Cauchy (12.0.2), hence convergent say $Az^{(m)} \to y$. Now $z^{(m)} \to z$ in ω_A since this is an AK space (4.3.8) hence $Az^{(m)} \to Az$ in

ω. But $Az^{(m)} \to y$ in ω since Y is an FK space hence $y = Az$ so $z \in S$ by 12.1.7.

8. EXAMPLE. Let $Y = \ell$, $(Ax)_n = x_n - x_{n-1}$. Then $\ell_A = bv$ which has $B = bv$, $S = W = F = F^+ = bv_0$.

From the results we can obtain *inequivalence theorems:*

9. EXAMPLE. *There exists no matrix* A *such that* $\ell_A = c$. This is immediate from Theorem 7 and the fact that for $X = c$, $F = c$, $W = c_0$. Compare Example 8. Also $\ell_A = c_0$ or ℓ^∞ is not possible since for c_0, ℓ^∞, $F^+ = \ell^\infty$, $W = c_0$.

10. EXAMPLE. Suppose that Y has AK and weakly convergent sequences are convergent, e.g. $Y = \ell$, bv_0. Then for any matrix A with columns in Y: $L_e^+ = L_e = L_a = S = W = F = F^+$ (Theorem 7, 12.3.2).

11. Recall the condition (**) if Y_A is sc then $Y_A \supset bv$ (i.e. Y_A is vsc), given in 9.5.11. There it was pointed out that $Y^{\beta f} = Y$ is sufficient. Another sufficient condition is that weakly convergent sequences in Y are convergent for this implies that $F^+ = F$ (Theorem 7) and so 10.2.7 (iii), (iv) show that (**) holds. Thus bv_0 also satisfies (**).

12. COROLLARY. *With* Y, A *as in Remark 1, if* Y_A *is sc it must be conull.*

For $1 \in F^+$ by 10.2.7, hence $1 \in W$ by Theorem 7.

13. EXAMPLE. *If* bv_A, $bv_0)_A$, *or* ℓ_A *is sc then* A *must be conull.* For each of these is included in c_A and is conull, hence c_A is conull (9.3.6). It is proved in [42] Theorem 3 that if A is a regular Hausdorff matrix then bv_A is variational. Here we see that it cannot be sc.

A space Y with the property of Remark 1 is in particular weakly sequentially complete. So also is any reflexive space. Any weakly sequentially complete space X obviously has $F^+ = F = W$. Suppose that $X \supset c_0$ as well. Then $X \supset F = F^+ \supset F^+(c_0) = \ell^\infty$, a bigness theorem. The inclusion must be proper since ℓ^∞ is not weakly sequentially complete (because $F = \ell^\infty$, $W = c_0$.)

Sources: [13], [86], [87].

12.5. c-LIKE SPACES

A *c-like space* is a BK space X such that X^β has AD and X is closed in $X^{\beta f}$. For example c_0, c, ℓ^∞, bv, bs are c-like spaces. Their study brings us close to summability.

1. If X is c-like then X is closed in $X^{\beta\beta}$ (4.2.5). Also if X^β has AK and X is closed in $X^{\beta\beta}$ then X is c-like by 7.2.7 (ii).

2. REMARK. In this section *Y is a c-like space and A a matrix such that* $Y_A \supset \phi$. *The distinguished subspaces are calculated for* Y_A.

3. THEOREM (J.J. Sember and G. Bennett.) *With the notation of Remark 2,* $L_e^+ = B^+ \cap \omega_A$, $L_e = L_a = B$.

Let $z \in L_e^+$. For $t \in Y^\beta$, $(tA)z = (A\cdot z)^T t$ so $(A\cdot z)^T \in (Y^\beta : cs)$. Hence $A\cdot z \in (bv : Y^{\beta f})$ by 8.3.8, and so $A\cdot z \in (bv_0 : Y^{\beta f})$. The columns of $A\cdot z$ are in Y, for they are multiples of the columns of A, so $A\cdot z \in (bv_0 : Y)$ by 8.3.6. Hence $z \in B^+$ by 12.1.3. The rest of the result is given in 12.2.3 and 12.2.5.

Example 12.2.7 shows that Theorem 3 is sharp.

4. EXAMPLE. Suppose Y is c-like and has AK e.g. $Y = c_0$.

Then for any matrix A with columns in Y, $L_e^+ = B^+ = F^+$, $L_e = L_a = B = F = W$. (Example 6 shows that S may be different.)

We shall apply Example 4 to $\Omega(r) = c_A^o$ (6.3.2).

5. EXAMPLE. *$\Omega(r)$ has AD.* It is sufficient, just as in 5.2.14, to show that the matrix A given in 6.3.2 is of type M. If $t \perp A$, a calculation yields $t_{r_i} - \sum \{t_k : r_i < k \leq r_{i+1}\} = 0$, $t_k = 0$ for $r_i < k < r_{i+1}$ for each $i = 1,2,\ldots$. Thus $t_{r_i} = t_{r_{i+1}}$ and so $t = 0$.

6. EXAMPLE. *$\Omega(r)$ is a conull conservative AD space with $B = F = W \neq S$. It has neither AK nor AB; and W, with its own BK topology is not an AD space.* Call the space X. Some of this is from Examples 4, 5. The absence of AK is because $1 \neq \sum \delta^k$ by 5.2.9. This shows also that $1 \notin S$; $1 \in W$ since X is conull, X does not have AB since this would imply AK by 10.3.19. Finally if W had AD it would follow that $S = W$ by 10.5.3. A more exact description is given in 13.4.15.

7. An amusing consequence is that since W (with its BK topology) does not have AK it does not have SAK by 10.5.1 i.e. $W(W) \neq W$. We shall see 13.2.26 that W is not even conull.

8. COROLLARY. *With the notation of Remark 2 these are equivalent: Y_A has AB, $A^T \in (Y^\beta : Y_A^\beta)$, $G \subset X^\beta$.*

This is immediate from Theorem 3, 12.2.8, 12.2.10. (Note that 8.3.8 says that $A^T \in (Y^\beta : Y_A^f)$ always.)

The best possible extension of 4.4.9 is now given. It is an extension since A conservative implies $B^+ \supset B^+(c)$ (10.2.9) $= \ell^\infty$. It is best possible by Theorem 3.

9. COROLLARY. *Let* A *be a matrix with convergent columns,* $f \in c_A'$. *Then there exists a sequence* γ *such that* $f(x) = \mu \lim_A x + \gamma x$ *for all* $x \in B$.

This follows from 4.4.3 with Theorem 3, $\gamma = tA + \alpha$.

The reader should note that most of the applications of Theorem 3 will be through this corollary.

CHAPTER 13
DISTINGUISHED SUBSPACES OF c_A

13.1. B AND THE PERFECT PART

1. Many results of earlier chapters assume a more definitive and instructive form when phrased in terms of distinguished subspaces.

2. Recall the set ϕ_1 (5.2.8) of eventually constant sequences. Its closure in X is called the *perfect part* of X and X is called *perfect* if this set is dense. This agrees with previous definitions (§3.3) since if X is conservative, $\overline{c} = \overline{\phi_1}$ by 4.2.7 which also implies that the perfect part is $\overline{bv}$ if $X \supset bv$, e.g. if X is vsc (9.6.1).

3. Let X be an FK space $\supset \phi$. The following facts are immediate from 10.2.7: $B^+ \supset \phi_1$ iff $X \supset bv_o$, $B \supset \phi_1$ iff $X \supset bv$, $F^+ \supset \phi_1$ iff X is sc, $F \supset \phi_1$ iff X is vsc, $W \supset \phi_1$ iff X is conull, $S \supset \phi_1$ iff X is strongly conull.

4. Numerous references will be made to 12.5.3 which says that $B = L_a = L_e$.

5. THEOREM. *Let A be coregular, $X = c_A$. Then the perfect part of X is equal to $\overline{B}$.*

Since $B \supset \phi_1$ (Remark 3) it is sufficient to show that $\overline{\phi}_1 \supset B$. Let $f \in X'$, $f = 0$ on ϕ_1. In particular $\chi(f) = 0$ so $\mu = 0$ by 9.6.8. For $x \in B$, $f(x) = \gamma x$ (12.5.9). Now $\gamma_k = f(\delta^k) = 0$ so $f = 0$ on B and the result follows by the Hahn-Banach theorem (3.0.1).

The perfect part may be larger than B *i.e.* B *need not be closed.* (10.3.22; but 13.4.10 is easier.) The situation for conull matrices is discussed in 13.5.1, 13.5.2.

Theorem 5 extends 4.6.7 (which says that the perfect part includes the bounded sequences) in two directions: first $B \supset c_A \cap \ell^\infty$ if A is conservative since $B^+ \supset B^+(c) = \ell^\infty$ (10.2.9). Second, A is allowed to be sc. There is an obvious extension of the consistency theorem 5.3.3: replace c by ϕ_1, $c_A \cap \ell^\infty$ by B.

6. DEFINITION. *A matrix* A *is said to have* AB *or is called an* AB *matrix if* c_A *has* AB.

7. COROLLARY. *A coregular* AB *matrix is perfect.*

By Theorem 5. The converse is false. (13.4.10).

Theorem 5.4.11 is now extended to sc matrices.

8. THEOREM. *Let* A *be coregular and reversible. Then* A *is perfect iff it is of Type* M.

Sufficiency: if $f = 0$ on ϕ_1, $f(x) = t(Ax)$ by 5.4.3 since $\mu = 0$ by 9.6.8. Also $0 = f(\delta^k) = (tA)_k$ so $t = 0$. Thus $f = 0$ and the Hahn-Banach theorem yields the result. Necessity: let $t \perp A$, $f(x) = t(Ax)$. Then $f = 0$ on B (Remark 4), hence on ϕ_1 (Remark 3) so $f = 0$. Fix n and choose x such that $Ax = \delta^n$. Then $0 = f(x) = t_n$ so $t = 0$.

13.2. THE INSET AND REPLACEABILITY

If A is a matrix with convergent columns i.e. $c_A \supset \phi$, we have had numerous encounters with the sequence $a = \{a_k\}$, $a_k = \lim a_{nk}$. If $A \in \Phi$ then $a \in \ell$ (1.3.7); if A is sc then $a \in cs$ (9.5.6). It is slightly paradoxical and very important that $x \in c_A$ does not imply $x \in a^\beta$:

1. EXAMPLE. Let $(Ax)_n = \sum_{k=1}^{2n} x_k$. Then $a = 1$ and $\{(-1)^n\} \in c_A \setminus a^\beta$. The example can easily be modified to give a conservative triangle by repeating each row, replacing the 1's by members of a sequence in ℓ, and replacing diagonal 0's with small numbers. There will be plenty of coregular examples also. (Example 10.)

2. REMARK. *In this section* A *is a matrix with convergent columns,* $a = \{a_k\}$ *where* $a_k = \lim a_{nk}$. *The distinguished subspaces are calculated in* $X = c_A$.

3. DEFINITION. *With the notation of Remark 2, In, the inset of* A, *is* $a^\beta \cap X$. *If* $In = X$, A *is said to have maximal inset.*

Thus a multiplicative matrix, or more generally, any matrix with null columns, has maximal inset. The matrix of Example 1 does not.

The crucial relations among W, F, B are shown by means of the inset.

4. LEMMA. *With the notation of Remark 2,* $F^+ \subset a^\beta$, $F \subset In$.

Let $f = \lim_A$, $z \in F^+$. Then $\sum f(\delta^k)z_k = az$ converges.

A similar argument shows $B^+ \subset a^\gamma$. This seems less significant.

5. THEOREM. *With the notation of Remark 2,* $F^+ = B^+ \cap \omega_A \cap a^\beta$, $F = B \cap In$.

Let $z \in B^+ \cap \omega_A \cap a^\beta$. By 12.1.5 (iv) we have to show that

$$\{z_k g(a^k)\} \in cs \text{ for each } g \in c' \text{ where } a^k \text{ is the kth column of } A \tag{1}$$

The first case is $g = \lim$ and (1) says $z \cdot a \in cs$ which is true since $z \in a^\beta$.

The second case is $g(y) = ty$, $t \in \ell$. A simple minded application of 12.1.3 (iv) will yield only bs in (1), but $\sum z_k g(a^k) = \sum_k \sum_n t_n a_{nk} z_k = (tA)z$ and this converges, verifying (1), since $z \in B^+ \cap \omega_A = L_e^+$. (12.5.3). Thus $B^+ \cap \omega_A \cap a^\beta \subset F^+$. But also $F^+ \subset B^+$, $F^+ \subset \omega_A$ (12.2.3), $F^+ \subset a^\beta$. (Lemma 4).

6. COROLLARY. *If* A *is replaceable,* $F = B$.

Let $c_A = c_M$ where M has null columns so that $\mathrm{In}(M) = c_M$. By Theorem 5, $F = B$ for c_M, hence for c_A since it is the same space.

In [71], it is pointed out that the converse of Corollary 6 is false, even for a coregular conservative matrix, also that the matrix of 5.2.5 has $F \neq B$.

7. EXAMPLE. *A non-replaceable conull conservative matrix.* The example can easily be modified to be a triangle by repeating rows and replacing zeros on the diagonal with a null sequence. Let $(Ax)_n = \sum_{k=1}^{n} (1/2)^{k-1}(x_{2k-1}+x_{2k})$. If $x = (1,-1,2,-2,3,-3,\ldots)$ then $Ax = 0$, $x \notin \mathrm{In}$ since $a = (1,1,\frac{1}{2},\frac{1}{2},1/3,1/3,\ldots)$ so that ax is divergent. Also $x \in B$ for if m is even, $Ax^{(m)} = 0$, while if m is odd, $(Ax^{(m)})_n = 0$ for $n < (m-1)/2$, 1 for $n \geq (m-1)/2$; so $\|Ax^{(m)}\|_\infty = 0$ or 1 for m even or odd. Thus $x \in B$ by 12.1.3. It has now been proved that $B \not\subset \mathrm{In}$, hence $B \neq F$ by Lemma 4 and the result follows by Corollary 6. (*This matrix is* μ-*unique* by Theorem 25.)

There is an intimate relationship between F and the inset. (Theorem 9). We first introduce some natural invariants:

8. DEFINITION. *Let* $U = U(A)$ *be any set associated with a space of the form* c_A. *Then* U^i, *the internal* U, *is* $\cap\{U(M): c_M = c_A\}$, U^e, *the external* U, *is* $\cup\{U(M): c_M = c_A\}$.

For example, the *internal inset*, $In^i = \cap\{In(M): c_M = c_A\}$ $= \cap\{m^\beta \cap c_M : c_M = c_A\} = c_A \cap(\cap\{m^\beta : c_M = c_A\}$, where $m = \{m_k\}$, $m_k = \lim m_{nk}$. The next result solves the naming problem for the internal inset.

9. THEOREM. *With the notation of Remark 2,* $F = In^i$.

$\subset$: By Lemma 4 and the invariance of F. $\supset$: Let $x \in In^i$. It is sufficient to show that $x \in B$ (Theorem 5) i.e. $x \in L_e$ (12.5.3). Let $t \in \ell$, form Mazur's matrix $M = M(t)$ (1.8.12), and set $D = MA$ so that $c_D = c_A$ by 1.6.4. Then $d_k = \lim_D \delta^k$ $= \lim_M a^k = \lim a^k + (tA)_k$ (1.8.12) $= a_k + (tA)_k$. Thus $tA = d - a$. Since $x \in d^\beta \cap a^\beta$ it follows that $(tA)x$ converges i.e. $x \in L_e$.

But there may be no minimum inset i.e. F need not be the inset of any equipotent matrix ([9] p. 229, Ex 2). Example 13 shows the opposite extreme.

10. EXAMPLE. *In is not invariant.* Let A be a regular matrix which does not have AB. (One is given in 13.4.10.) It has maximal inset since its columns are null. However it does not have AB, a fortiori $F \neq c_A$. By Theorem 9 there is an equipotent matrix with smaller inset. It is interesting that in spite of this, *$B \cap In$ is invariant* since it is F. (Theorem 5).

Of course if c_A has FAK then A has *PMI* (Propagation of Maximal Inset) i.e. every equipotent matrix has maximal inset, by Theorem 9. The ingenious example given next shows a matrix which does not have PMI but has invariant inset i.e. $In(M) = In(A)$ whenever $c_M = c_A$. A few results are given in preparation:

11. LEMMA. *If* $z \in bv \backslash bv_o$ *then* $z^\beta \subset cs$.

Since $\lim z \neq 0$ we can assume $z_i \neq 0$ for all i since changing finitely many terms of z doesn't affect anything. Let $x \in z^\beta$ then $x = (x \cdot z) \cdot (1/z) \in cs \cdot bv \subset cs$ by Abel's test. (That $1/z \in bv$ is because $|z_{n+1}^{-1} - z_n^{-1}| = |z_n - z_{n+1}|/|z_n z_{n+1}| < M|z_n - z_{n+1}|$.)

12. LEMMA. *Let* a, b *be sequences with* $a_i \neq 0$ *for all* i *and assume that* $b^\beta \supset a^\beta$ *properly. Then* $b^\beta \supset a^\gamma$.

Let $z = b/a$. Then $z^\beta \supset 1^\beta = cs$ properly. Thus $z \in z^{\beta\beta} \subset bv$. By Lemma 11, $z \in bv_o$ so $z^\beta \supset bs = 1^\gamma$ from which the result follows.

13. EXAMPLE (G. Bennett). *A matrix with invariant inset which is not maximal.* Let A be the matrix of Example 7; it was shown then that A does not have maximal inset. First, $B^+ = a^\gamma$ since $(Az^{(r)})_n = \sum_{k=1}^{s} a_k z_k$ for suitable s, so $F^+ = a^\beta$ by Theorem 5, hence $F = a^\beta$ since obviously $a^\beta \subset c_A$. Now if $c_M = c_A$ with $In(M) \neq In(A)$ we must have, using Lemma 4, $m^\beta \supset In(M) \supset F = a^\beta = In(A)$ so $m^\beta \supset a^\beta$ properly. By Lemma 12, $m^\beta \supset a^\gamma = B^+$ so $In(M) \supset B$. By Theorem 5 this implies $F = B$ contradicting the result of Example 7.

The inset is the natural domain of the function $\Lambda(x) = \lim_A x - ax$ which was defined for a special domain in 5.6.3. If A is vsc then $\Lambda(1) = \chi(A)$.

14. THEOREM. *With the notation of Remark 2,* $W = F \cap \Lambda^\perp$ *so either* $W = F$ *or* W *is a maximal subspace of* F. *If* A *is coregular* $F = W \oplus 1$.

If $x \in W$ the definition of W applied to $\lim_A$ shows that $\Lambda(x) = 0$. Conversely if $x \in F \cap \Lambda^\perp$, $f \in c_A'$, then $f(x) = \mu \lim_A x + \gamma x$ (12.5.9) $= \mu ax + \gamma x = \beta x$, say. Then $\beta_k = f(\delta^k)$ so $x \in W$.

15. With the notation of Remark 2, $a \in X^f$ since $a_k = \lim_A \delta^k$. (We saw in Example 1 that $a \notin X^\beta$ is possible.) Thus Λ is X-continuous on W (11.1.2) but actually $\Lambda = 0$ on W by Theorem 14.

16. THEOREM. *With the notation of Remark 2 suppose that A is coregular (vsc). Then these are equivalent (i) A is replaceable, (ii)* $1 \notin \overline{\phi}$ *(closure in X), (iii) lim is X-continuous on* ϕ_1*, (iv)* Λ *is X-continuous on* ϕ_1*, (v)* $x \to ax$ *is X-continuous on* ϕ_1*, (vi) In (iv), (v) replace* ϕ_1 *by F.*

This is the extension of 5.2.1 to sc matrices.

(iv) = (v): $\Lambda(x) = \lim_A x - ax$ and $\lim_A \in X'$. The two parts of (vi) are equivalent for the same reason.

(i) implies (iii): Let $c_M = c_A$, M having null columns, $\chi(M) = 1$. Then $\lim_M = 0$ on ϕ, $\lim_M 1 = 1$ so $\lim_M = \lim$ on ϕ_1 and continuity is guaranteed by invariance of topology (4.2.4).

(iii) implies (i): Let $f \in X'$, $f = \lim$ on ϕ_1 by the Hahn-Banach theorem (3.0.1). Then $1 = \chi(f) = \mu\, \chi(A)$ (9.6.8) so $\mu \neq 0$. The result follows by 5.1.1.

(iii) implies (ii): Since $\lim = 0$ on ϕ and $\lim 1 = 1$.

(ii) implies (iii): In ϕ_1 with the X topology, ϕ is a maximal subspace which is not dense — hence it is closed (5.0.1). Since it is $\lim^\perp$, lim is continuous (5.0.1).

(iii) = (iv): For $x \in \text{In}$, $\Lambda(x) = \lim_A x - ax$. This holds in particular for $x \in \phi_1$ since $a \in cs$ (9.5.6). Thus $\Lambda(1) = \chi(A) = \chi(A) \lim 1$ and for $x \in \phi$, $\Lambda(x) = ax - ax = 0 = \chi(A) \lim x$. Thus $\Lambda = \chi \cdot \lim$ on ϕ_1 and the result follows since $\chi \neq 0$.

(vi) implies (iv) since $F \supset \phi_1$ (13.1.3).

(ii) implies (vi): In F with the X topology, W is a maximal subspace (Theorem 14) which is not dense since $W \subset \overline{\phi}$ (10.2.6) and $1 \in F$ (10.2.7). Hence W is closed (5.0.1). But $W = \Lambda^{\perp}$ (Theorem 14) and so Λ is continuous (5.0.1).

17. If A is coregular and non-replaceable it follows from Theorem 16 that ϕ_1, F, W, In are all non-closed subspaces of $X = c_A$, for, by the Banach-Steinhaus theorem (1.0.4), Λ is continuous on any closed subspace of its domain. If W is closed, it is closed in F with the X topology and as proved above this makes Λ continuous. Of course Λ is always continuous on F with its FK topology (10.4.10, 10.3.26).

18. THEOREM. *With the notation of Remark 2, suppose that A is non-replaceable. Then (i) A is μ-unique, (ii) if $f \in X'$, $f = 0$ on ϕ, then $\mu = 0$ in every representation of f, (iii) ϕ is dense in B (with the topology of X).*

With f as in (ii) suppose that f has a representation as in 4.4.3 with $\mu \neq 0$. By 5.1.1 there is an equipotent matrix M with $\lim_M = f$. Then M has null columns since $f(\delta^k) = 0$, so A is replaceable. This proves (ii), and (i) follows. To prove (iii) let $f = 0$ on ϕ. By (ii) and 12.5.9, $f(x) = \gamma x$ for $x \in B$. But $\gamma_k = f(\delta^k) = 0$ so $f = 0$ on B, and the result follows from the Hahn-Banach theorem (3.0.1).

19. COROLLARY. *With the notation of Remark 2, $\overline{F} = \overline{B}$. If either B or F is closed then $F = B$.*

Since $\phi \subset F \subset B$, the first part follows for a non-replaceable matrix from Theorem 18 (iii). If A is replaceable, $F = B$. (Corollary 6). If B is closed, so also is F (10.4.9) and the last result follows.

The next result shows that the only role of the assumption of null columns is to ensure maximal inset; for replaceablity it is enough to have an equipotent matrix with maximal inset. An amusing corollary is that if A is given an extra row, namely a, then the resulting matrix, call it M, has maximal inset, hence is replaceable. Also $c_M = c_A \cap a^\beta$ is regular if A is coregular.

20. THEOREM. *Any matrix with maximal inset is replaceable.*

For Λ is defined on c_A, hence continuous by the Banach-Steinhaus theorem 1.0.4. The result follows by 5.1.1.

21. *If* c_A *has closed inset,* $F = B$. $[B \subset \overline{B} = \overline{F} = \overline{B \cap In} \subset \overline{In} = In]$. *If* A *is also coregular it is replaceable* by Remark 17. I do not know whether A must be replaceable in general. See 13.5.5.

22. EXAMPLE. Λ *continuous but* In *not closed,* A *a coregular replaceable triangle.* With Q as in 1.2.5, $X = c_Q$, let $u \in X^f \backslash X^\beta$ (11.1.7, 11.1.8). Now $ux = t(Qx)$ for $x \in u^\beta \cap X$ (11.1.7); let $f(x) = \lim_Q x + t(Qx)$. There exists a triangle A such that $c_A = X$, $\lim_A = f$; this is by 5.1.1 (with A replaced by Q) in the proof of which D is a triangle and there is no need to continue since $\alpha = 0$ (3.2.4, 1.7.15). Then $a_k = f(\delta^k) = (tQ)_k = u_k$ as in 11.1.7. So $In(A) = u^\beta \cap X$ which is not closed in X (11.1.7). However $\Lambda(x) = \lim_A x - ux$ is continuous on In (11.1.7).

23. EXAMPLE (G. Bennett). $\Lambda = 0$ *on* In *but* In *not closed.* The matrix of Example 7 has $F \neq B$ so In is not closed by Remark 21. If $x \in In$, $(Ax)_m = \sum_{k=1}^{2m} a_k x_k \to ax$ so $\Lambda = 0$ on In. Note also that $In(M)$ is not closed for any equipotent matrix M since $F \neq B$ is an invariant condition. In general *it is possible for* In *to be closed in* c_A *and not in* c_M *with* $c_A = c_M$*:* for example if A is perfect, regular and not AB (13.4.10), $In = c_A$

but there exists M with $c_M = c_A$, $In \neq c_M$ (Theorem 9); thus $In \supset c$ so it is dense.

24. EXAMPLE. The matrix of Example 7 has invariant inset (Example 13) and $\Lambda = 0$ (Example 23), so $W = F = In$ (Theorems 9 and 14). Any easy calculation shows $S = In$. (This also follows from [87] 8.2: if $\Lambda^{\perp} = W$ then $W = S$.) Thus *$W = S$ does not imply that W is closed.* (Example 23.) Compare Theorem 10.5.2.

25. THEOREM. *With the notation of Remark 2 suppose that A is not μ-unique. Then A is replaceable and $W = F = B = \overline{\Phi}$.*

This is just a summary of Corollary 6, Theorems 14 and 18.

26. EXAMPLE. Let A be any conull conservative triangle with $B = c_A \cap \ell^{\infty}$. (There are plenty of these — see 13.4.9.) Then $c \subset W \subset \ell^{\infty}$ so W is coregular and conservative in its own BK topology (10.5.5, 4.6.3).

27. There is a subtle commentary on the use of abstract methods available here. First note that $W(A) = \cap\{\Lambda_M^{\perp}: c_M = c_A\}$. This is proved in the same way as Theorem 9. Agnew and Zeller showed that if A is conull and conservative, there exists r such that $\Omega(r) \subset c_A$. (Generalized in 6.4.3.) They showed more: $\Omega(r) \subset \Lambda^{\perp}$. This implies that the sequence r chosen depends on A and cannot be defined invariantly from the FK structure of c_A as in 6.4.3; for $\Omega(r) \not\subset W$ if A is as in Example 26 since W is coregular (6.4.4), and so the formula for W given a moment ago shows that there exists M with $c_M = c_A$, $\Omega(r) \not\subset \Lambda_M^{\perp}$.

Sources and further results: [9], [16], [44].

13.3. THE MAIN THEOREM

1. THEOREM. *Let* A *be a matrix with convergent columns,* $X = c_A$. *These are equivalent:*

(i) X *has* AB, $B = X$,

(ii) X *has* FAK, $F = X$,

(iii) $L_e = X$ *(Definition 12.2.2 with* $Y = c$*)*,

(iv) A *is associative i.e.* $L_a = X$,

(v) A *has PMI (Text following 13.2.10)*,

(vi) X *has* AK *or* $X = S \oplus u$ *with* $u \in B$ *i.e.* X *has* $\{u, \delta^n\}$ *as basis,*

(vii) $W = X$ *or* $X = W \oplus u$ *with* $u \in B$,

(viii) $A^T \in (\ell : X^\beta)$.

If A *is coregular replace (vi), (vii) by*

(vi)' $X = S \oplus 1$ *i.e.* X *has* $\{1, \delta^n\}$ *as basis,*

(vii)' $X = W \oplus 1$.

(i) = (iii) = (iv) by 12.5.3; (ii) = (v) by 13.2.9; (i) = (ii): If $B = X$, $F = B = X$ by 13.2.19. The converse is trivial since $F \subset B$; (vi) implies (vii) and (vi)' implies (vii)' since $S \subset W$; (vii) implies (i) since $W \subset B$; (ii) implies (vi): Note that Λ is defined on F (13.2.4) hence $\Lambda \in X'$ by the Banach-Steinhaus theorem (1.0.4) and so W is either a closed maximal subspace or $W = X$ since $W = F \cap \Lambda^\perp$ (13.2.14). By 10.5.2, S is closed so $S = W$ since $\phi \subset S \subset W \subset \overline{\phi}$ (10.2.6); (iii) = (viii) by 12.2.8.

If A is coregular, $1 \in F \setminus W$ by 10.2.7 so the last part is a special case.

Other conditions equivalent to the important AB conditions are given in 14.5.3.

2. Some easy inequivalence theorems follow, namely, there exists no matrix A such that $c_A = cs \oplus u$, $u \in bs$, for the latter

space satisfies (vi) but not (ii) by 10.4.12. Similarly $c_A \neq bv$ (10.4.5). The last result can be phrased: *if* $c_A \supset bv_o$ *properly (e.g. if* A *is sc), then* A *must sum a sequence which is not in* bv — a bigness theorem.

In (vi), (vii), $u \in B$ may be replaced by $u \in F$ obviously. However, it may not be replaced by $u \in In$ even if A is a multiplicative triangle and all the distinguished subspaces are closed and equal. (The last condition is bravado — see Remark 4.):

3. EXAMPLE. *A multiplicative matrix* A *such that* $c_A = S \oplus u$ *but* A *does not have* AB. It is sufficient to construct A with these properties but with "multiplicative" replaced by "conservative and replaceable". Let $A = DE$ where $(Dx)_n = x_n - 2x_{n-1}$, $(Ex)_n = \sum_{k=1}^{n} e_k y_k$ with $e \in \ell$, $e_k \neq 0$. Then A is a conservative triangle and $c_A = c_E \oplus u$ where $u = E^{-1}\{2^{-n}\}$ (1.8.8). Now $c_E = e^\beta$ has AK (4.3.7) and is a closed subspace of c_A (4.5.5) hence is equal to S. Also A is not of type M since with $t_n = 2^{-n}$ we have $t \perp D$ hence $t \perp A$, thus A does not have AB by 13.1.8, 13.1.7. Finally we have to show that A is replaceable. Let $(Mx)_n = \sum_{k=n+1}^{\infty} e_k y_k$. Then $c_M = c_E = e^\beta$ and so $c_A = c_{DM}$ since D is a triangle. Clearly DM has null columns. (By the next Remark, $B = F = W = S$.)

4. *If the distinguished subspaces of* $X = c_A$ *are not all equal and* $X = S \oplus u$ *then* A *has* AB. *In particular the implication holds if* A *is coregular.* Since S is closed, $S = W$ so $F = S$ or X by 13.2.14. If $F = S$ then $B \neq F$ so $B = X$; if $F = X$ the same is true.

13.4. APPLICATIONS

Obviously Theorem 13.3.1 is very powerful. It shows that an AB matrix has a convergence domain which is either an AK space (like c_0) or an AK space with one more sequence (like c). In either case c_A has a basis of very special form. For coregular A the basis is $\{\delta^k\} \cup \{1\}$, exactly the same as that of c. In order to show applications we shall give growth theorems for B and S, and a criterion for AB which will identify matrices falling within the scope of the applications.

1. REMARK. In this section *A is a matrix with convergent columns, $X = c_A$ and the distinguished subspaces are calculated in X.*

2. THEOREM. *With the notation of Remark 1, let $u_k = \sup_n |a_{nk}| = \|a^k\|_\infty$. Then each $x \in S$ satisfies $u_k x_k \to 0$; in particular $a_{kk}x_k \to 0$.*

By 12.1.7, $A \cdot x$ is strongly conull. By 5.2.9, $\sum a_{nk}x_k$ is uniformly convergent. Let $\varepsilon > 0$ and choose k_0 such that $|\sum_{k=r}^{\infty} a_{nk}x_k| < \varepsilon/2$ for all n if $r > k_0$. Then $|a_{nk}x_k| = |\sum_{i=k}^{\infty} - \sum_{i=k+1}^{\infty} a_{ni}x_i| < \varepsilon$ for $k > k_0$. Taking $\sup_n$ gives the result.

3. EXAMPLE. With the notation of Remark 1, *if A is conservative and each column has a 1 in it (or more generally, $u_k > \varepsilon$ in Theorem 2) then $S = c_o$.* For $S \supset S(c) = c_0$ (10.2.9), and conversely by Theorem 2.

APPLICATION I. *A best possible growth theorem.* (That it is best possible will be clear since the columns of $A^{-1} \in c_A$.):

4. COROLLARY. *Let* A *be a coregular matrix with* AB *and suppose that* $a_{nn} \to 0$. *Then each* $x \in c_A$ *satisfies* $x_n = o(1/a_{nn})$.

For $x \in S$ this follows from Theorem 2. For $x = 1$ it is obvious. But this is all there is by 13.3.1, (vi)'.

5. EXAMPLE. Since $(C,1)$ has AB (it has a monotone norm by 7.1.6, and has AB by 10.3.12) this yields the result of 3.3.11.

6. COROLLARY. *Let* A *be a conservative matrix with* AB *and suppose that* $\{1/a_{nn}\}$ *is bounded. Then* A *is Mercerian.*

First $S = c_0$ by Example 3. Now S is at least a maximal subspace of $X = c_A$ (13.3.1). Since $1 \in X \backslash S$ it follows that $X = S \oplus 1 = c$.

APPLICATION II. *A proof of 4.6.8 for triangles.* (The proof can easily be adapted to apply in general.):

7. COROLLARY. *Let* A *be a conservative triangle which sums only bounded sequences. Then* A *is Mercerian.*

For A has AB since $B^+ \supset B^+(c) = \ell^\infty$ (10.2.9) and $\{1/a_{nn}\}$ is bounded by 1.7.7. The result follows by Corollary 6.

8. THEOREM. *With the notation of Theorem 2, each* $z \in B^+$ *satisfies* $u \cdot z \in \ell^\infty$; *in particular* $\{a_{nn}z_n\}$ *is bounded.*

This is the special case of 10.3.1 in which $p(x) = \|x\|_A$.

Thus, for example, *if* A *has no zero columns then* B^+ *has a growth sequence.*

9. EXAMPLE. *If* A *satisfies the conditions of Example 3 then* $B^+ = \ell^\infty$, $B = c_A \cap \ell^\infty$. For $B^+ \supset B^+(c) = \ell^\infty$ (10.2.9) and conversely by Theorem 8.

10. EXAMPLE. The matrix Q (1.2.5) is a perfect (Type M)

regular triangle (3.3.8); $B = c_Q \cap \ell^\infty$ (Example 9) hence B is dense in $X = c_Q$; B is a proper subset of X, i.e. Q does not have AB, by Corollary 6 and the fact that $\{(-1)^n\} \in X$. (This is also immediate from Theorem 14.) Thus finally, *B is not closed.*

In order to apply these results one needs methods to recognize AB matrices. A sufficient condition will now be given. Others may be found in [84].

11. EXAMPLE. (Bosanquet's criterion). *Let A be a triangle with convergent columns, $d_{nk} = a_{n+1,k}/a_{nk}$ for $k \le n$. Suppose that $0 \le d_{nk} \le 1$ and that d_{nk} is decreasing (not strictly) as a function of k for each n. Then A has AB.* Fix $x \in X = c_A$. We shall show that for all m, $\|x^{(m)}\| \le \|x\|$, i.e. for every n

$$|(Ax^{(m)})_n| \le \|x\| \quad \text{for all } m. \tag{1}$$

This is trivial for $n = 1$ since $|(Ax^{(m)})_1| = |(Ax)_1| \le \|x\|$. To prove it by induction on n we consider $|(Ax^{(m)})_{n+1}| = |(Ax)_{n+1}| \le \|x\|$ if $n + 1 \le m$ while if $n + 1 > m$ the expression is $|\sum_{k=1}^{m} a_{n+1,k}x_k| = |\sum d_{nk}a_{nk}x_k| \le \max\{|\sum_{k=1}^{r} a_{nk}x_k| : 1 \le r \le m\}$ (Abel's inequality 1.2.10) $\le \|x\|$ by the induction hypothesis that (1) holds for n.

The (C,1) matrix has $d_{nk} = n/(n+1)$. This satisfies the criterion of Example 11, hence is an AB matrix. We already knew a little more as pointed out in Example 5. The methods of Example 5 can be applied to any matrix which satisfies Bosanquet's criterion.

APPLICATION III. *Mazur's matrix* (1.8.12). For simplicity assume $0 < t_i \le 1$. The matrix is Mercerian by Example 11 and

Corollary 6.

APPLICATION IV. *Mercer's theorem* (2.4.1). A form of this is that $A = (C,1) + \alpha I$ is Mercerian for $\alpha > 0$. Here $d_{nk} = n/(n+1)$ for $k < n$, $n/(n+1)(n\alpha+1)$ for $k = n$. The result follows by Example 11 and Corollary 6.

We now give some conditions for AB triangles. The classical theory dealt mainly with triangles and these results were extensively used.

12. THEOREM. *Let A be a triangle with convergent columns, $X = c_A$. Then A has AB iff there exists M such that*

$$\left|\sum_{k=1}^{m} a_{nk}x_k\right| \leq M \max\{|(Ax)_i| : 1 \leq i \leq m\}$$

for all m, n, x. (Note that $x \in X$ is not required.)

Necessity: Define $u_m: X \to c$ by $u_m(x) = Ax^{(m)}$. Then $\{u_m\}$ is pointwise, hence uniformly, bounded (1.0.3). Let $M = \sup\|u_m\|$. Then for all m, n and any sequence x, $|A(x^{(m)})_n| \leq \|u(x^{(m)})\| \leq M\|x^{(m)}\|$ which gives the result.

Sufficiency: Trivial since for $x \in X$ the first term is $|(Ax^{(m)})_n|$ and so $\|x^{(m)}\|$ is bounded.

By taking $y = Ax$, $z = A^{-1}y^{(m)}$ (m fixed) so that $z \in X$ it is easy to calculate that Theorem 12 holds with $M\|x\|_A$ for $x \in X$ on the right side of the inequality.

13. COROLLARY. *Let A be a triangle with AB. Then there exists M such that $|a_{nk}| \leq M|a_{kk}|$ for all n,k.*

Taking $x = \delta^m$ in Theorem 12 yields $|a_{nm}| \leq M|a_{mm}|$.

This result shows that the substitution of a_{kk} for u_k in Theorems 2 and 8 and elsewhere is not a genuine improvement.

Corollary 13 of course supplies easy examples of non-AB matrices. An improvement now given will make this even easier as the subsequent examples show. For one thing there is no need to involve Mercerian considerations.

14. THEOREM. *Let* A *be a regular triangle with* AB. *Then* $\sum |a_{nn}| = \infty$.

If the conclusion is false, fix m. Then for $n > m$, $|(A1)_n| \leq U_n + V_n$ where $U_n = \sum_{k=1}^{m} |a_{nk}| \to 0$ as $n \to \infty$ and

$$V_n = \sum_{k=m+1}^{n} |a_{nk}| \leq M \sum_{k=m+1}^{\infty} |a_{kk}| \quad \text{(Corollary 13)}$$

which can be made arbitrarily small. Thus $\lim_A 1 = 0$.

As an application we get that *the square of an AB triangle need not have AB*, indeed if $A = (C,1)^2 = H^2$, $\sum |a_{nn}| = \sum 1/n^2$. ($(C,1)$ has AB as pointed out after Example 11.)

Indeed all the Cesaro and Hölder matrices of order greater than 1 fail to have AB by the same reasoning.

15. EXAMPLE. Some more information about $\Omega(r)$ is available. Clearly Theorem 8 applies to ℓ_A^∞ and $(c_o)_A$. So Theorem 8 and Example 3 with 12.5.6 yield $B = F = W = \Omega_b(r)$, $S = c_o$.

16. EXAMPLE. *(i) does not imply (ii) in 9.6.5.* Let M be a matrix such that $Z = c_M$ has AD but not AB. (For example let M be the coregular non-replaceable triangle of 5.2.5. Absence of AB is guaranteed by 13.2.20, 13.3.1 (v).). Then $t(Mx)$ exists for all $t \in \ell$, $x \in Z$, but $(tM)x$ fails to exist for some $t \in \ell$, $x \in Z$ by 13.3.1 (iii). Interchanging t, x and setting $A = M^T$ gives: *$(tA)x$ exists for all $x \in \ell$, $t \in Z$ but $t(Ax)$ fails to exist for some $x \in \ell$, $t \in Z$.*

History and other applications may be found in [39], [41], [58], [84], and the references given there.

13.5. ALMOST COREGULAR AND VERY CONULL

Conull spaces show a wider variety of behavior than do coregular spaces. We have already seen strongly conull spaces, those for which $1 \in S$ (5.2.11, 5.2.13, 10.2.7).

Here we discuss a classification which is suggested by two examples.

1. EXAMPLE. *A conull matrix with* $W \neq F$. Let $(Ax)_n = (-1)^n(x_n - x_{n-1})$ (Convention: $x_o = 0$). Then A is a multiplicative 0 triangle with $F = B = X \cap \ell^\infty$ where $X = c_A$. (13.4.9, 13.2.6). Also $\lim_A = 0$ on c but $\lim_A (-1)^n = 2$ so *the perfect part* $\supset B$. Moreover $\{(-1)^n\} \in F\backslash W$ by 13.2.14. *This matrix is* μ*-unique* by 13.2.25.

2. EXAMPLE. *A conull matrix with* $W = F$. Let $(Ax)_n = x_n - x_{n-1}$. (This is a special case of the matrix given in 6.3.2 — take $r_n = n$.) As in Example 1, $F = B = X \cap \ell^\infty$. Let $f(x) = \mu \lim_A x + t(Ax)$. If $f = 0$ on ϕ we have $0 = f(\delta^k) = \mu a_k + (tA)_k = (tA)_k = t_k - t_{k+1}$. Since $t \in \ell$ it follows that $t = 0$ so $f = \mu \lim_A$. By the Hahn-Banach theorem (3.0.1) it follows that $\overline{\phi} = (c_o)_A$. Now $\lim_A x = 0$ for every bounded $x \in X$ (1.7.11) so the perfect part $\supset B$. That $W = F$ follows from 13.2.14. (That *this matrix is* μ*-unique* is shown in 15.2.9.)

The matrix of 4.4.4 also has $W = F$ (indeed $c_A = a^\beta$ so $S = c_A$). However it is not μ-unique.

3. DEFINITION. *A vsc space is called almost coregular if* $W \neq F$, *very conull if* $W = F$. *The adjectives are also applied to*

a matrix A *according to which property applies to* c_A.

The preceding examples show that conull matrices of each type exist. A coregular matrix is almost coregular $(1 \in F\backslash W)$ and almost coregular matrices have many of the properties of coregular ones.

4. THEOREM. *A is almost coregular iff there exists x (necessarily in F) such that $A \cdot x$ is coregular.*

By 12.1.5, 12.1.6 with $Y = c$.

5. THEOREM. *An almost coregular matrix is μ-unique (compare 9.6.10). If it has closed inset it must be replaceable (compare 13.2.21). If it is reversible $b = 0$ in 5.4.5 (compare 5.4.6). If $c_A = S \oplus u$ then A has AB (compare 13.3.3).*

The first part is by 13.2.25. (The converse fails by Example 2).

If A has closed inset Λ is continuous by the Banach-Steinhaus theorem (1.0.4) so W is X-closed in F where $X = c_A$ (13.2.14). But $F = B$ (13.2.21) so W is X-closed in B. This implies that A is replaceable by 13.2.18 (iii).

That $b = 0$ follows from the μ-uniqueness and 5.2.6.

The last part is immediate from 13.3.4.

The reader is invited to scan the text for theorems with coregular in their hypothesis. This is a rich source of problems. Here are two negative results:

6. EXAMPLE. *A coercive almost coregular triangle.* (Compare 3.5.5). Let $(Mx)_n = x_n/n$, $x = \{n\}$. The $M \cdot x = I$ so $x \in F\backslash W$ by 12.1.5, 12.1.6 with $A = M$, $Y = c$.

7. EXAMPLE. *$M \supset A$ (1.7.9) with A very conull, M almost coregular* (compare 3.5.4). Let A, M be the matrices called A, B in 1.7.11 and apply Examples 2, 6. Note also that *CA may be almost coregular with C a regular triangle and A a very conull triangle* namely $C = (C,1)$ in which case $CA = M$.

Two other classifications of conservative conull matrices were introduced by E. Jurimae for the purpose of extending the Mazur-Orlicz bounded consistency theorem 5.6.11. A *conservative* matrix A is called a *J matrix* if $\bar{c} \supset b$ (where, in this section b stands for $c_A \cap \ell^\infty$), an *0 matrix* if $c_M \supset b$ and $\lim_M = \lim_A$ on c implies the latter equality on b. A coregular matrix has both of these properties (4.6.7, 5.6.11). Proof of the following facts may be found in [9], [22A]: *A conull matrix is an 0 matrix iff $\Lambda^\perp \supset b$.* (Hence *a replaceable very conull matrix is an 0 matrix* since $W = F \cap \Lambda^\perp$ and $F = B \supset b$, but not conversely by Example 6.) *Every 0 matrix is a J matrix and the converse holds for replaceable matrices but not in general. Every non-replaceable matrix is a J matrix.*

Sources: [7], [9], [16], [75A], [82].

CHAPTER 14
THE FUNCTIONAL μ

14.0. FUNCTIONAL ANALYSIS

1. The sum of two FH (or BH) spaces is an FH (or BH) space [76], pp. 39, 62. [80], Example 13-4-5.

2. Let X be a Banach space and f a linear functional. Then $f^{\perp}$ can be given a complete norm smaller than the restriction of the norm of X. Proof: Let $f(u) = 1$. Let $g \in X'$, $g(u) = 1$. Define $q: f^{\perp} \to g^{\perp}$ by $q(x) = x - g(x)u$ and take $\|q(x)\|$ as the new norm of x. (If $q(x) = 0$, $0 = f[x-g(x)u] = -g(x)$ so $x = 0$.)

3. A linear functional which is bounded on some neighborhood of 0 must be continuous. [80] Theorem 4-5-9.

4. Let X be a locally convex space. For $E \subset X$,

$$E^{\circ} = \{f \in X' : |f(x)| \le 1 \text{ for all } x \in E\}.$$

The set of all E° with E a bounded set in X (8.0.1) is a local base of neighborhoods of 0 for the *strong topology* on X'. If X is a normed space, the strong topology is given by the norm (1.0.1) [80], p. 119.

5. Let X, Y be locally convex spaces and $T: X \to Y$ linear and continuous. Define $T': Y' \to X'$ by $T'(g) = g \circ T$. This is the *dual map*. It is continuous when Y' and X' have their strong topologies. A special case is $X \subset Y$ (with the same topology) and T = inclusion. Then $T'(g) = g|X$ [80], Example 11-2-3.

6. Let X be a locally convex space, Y a vector space and $q: X \to Y$ a linear map such that $q^{\perp}$ is closed. Then Y has a locally convex topology such that q is continuous and a set $E \subset Y$ is closed iff $q^{-1}[E]$ is closed in X. This quotient topology is the largest which makes q continuous. [80], §6-2.

14.1 PARTS OF THE DUAL

One of the most intriguing problems in our subject is to decide whether the function μ, (14.2.2), is continuous. Apart from the intrinsic interest in a problem which is so easy to state is the importance of the function μ in connection with the fifth distinguished subspace P (Chapter 15). It also appears in results in which its role is unexpected, e.g. 14.5.3.

1. With $X = c_A$ recall $G \subset X'$ consisting of the set of f with the form $f(x) = t(Ax)$, $t \in \ell$; also the embedding of X^{β} in X' in which u corresponds to $\hat{u}$ with $\hat{u}(x) = ux$ (7.2.9).

2. EXAMPLE. If $X = c$ (i.e. $A = I$), then $G = X^{\beta} = \ell$, $X' = \ell \oplus \lim$ in the sense that each $f \in X'$ is $f(x) = \mu \lim x + tx$ (1.0.2).

3. EXAMPLE. *Let A be μ-unique and either reversible, or row-finite and one to one. Then $X^{\beta} \subset G$ (Hence A has AB iff $X^{\beta} = G$.)* Note that every coregular matrix is μ-unique (9.6.10), indeed every almost coregular matrix is (13.5.5). Let $u \in X^{\beta}$. Then $\hat{u}(x) = \mu \lim_A x + t(Ax)$ (5.4.3, 5.5.3), and $\mu = 0$ since this is true for $\hat{u}$ by definition. Thus $\hat{u} \in G$. (The last part is by 12.5.8.)

4. REMARK. In this section *A is a matrix and $X = c_A$ is assumed to be a BK space.*

5. THEOREM. *With the notation of Remark 4, the closure of* X^β *includes* G.

The topology involved is the norm topology of X'. Let $f(x) = t(Ax)$, $\varepsilon > 0$. Choose m so that $\sum_{n=m+1}^{\infty} |t_n| < \varepsilon$ and let $u_k = \sum_{n=1}^{m} t_n a_{nk}$. Obviously $u \in X^\beta$. For $x \in X$ with $\|x\| \leq 1$ we have

$$|f(x)-\hat{u}(x)| = |\sum t_n(Ax)_n - \sum u_k x_k| = |\sum_{n=m+1}^{\infty} t_n(Ax)_n| \leq \|Ax\|_\infty \sum |t_n|$$
$$< \varepsilon$$

since $\|Ax\|_\infty = \|x\| \leq 1$. Thus $\|f-\hat{u}\| \leq \varepsilon$

6. LEMMA. *With the notation of Remark 4,* G *can be made into a Banach space whose topology is larger than the* X'*-topology on* G.

Let $q\colon \ell \to X'$ be defined by $q(t)(x) = t(Ax)$. Then q maps ℓ onto G. Further the null-space of q is closed, for it is $\bigcap\{E(x): x \in X\}$ where $E(x) = \{t: t(Ax) = 0\}$ and each $E(x)$ is closed since $Ax \in c$ so that the map $t \to t(Ax)$ is a continuous functional on ℓ. Thus the quotient topology (7.0.1) makes G into a Banach space. To show that the quotient topology is larger than the X'-topology it is sufficient (14.0.6) to show that $q\colon \ell \to X'$ is continuous. Now $|q(t)(x)| = |t(Ax)| \leq \|t\|\cdot\|x\|_A$ hence $\|q(t)\| \leq \|t\|$ as required.

7. That this norm makes G a Banach space is true, with the same proof, without any restriction on A. However one must not expect too much, for example if $A = 0$ then $G = \{0\}$.

8. LEMMA. *With the notation of Remark 4,* $G + X^\beta$ *can be made into a Banach space whose topology is larger than the relative topology of* X'.

Here we are identifying X^β with its image in X' shown in Remark 1. With $H = X'$, G and X^β are BH spaces by Lemma 6 and 10.1.4. The result follows by 14.0.1.

9. LEMMA. *The same as Lemma 8 with larger replaced by smaller.*

First, $G + X^\beta = X'$ if A is not μ-unique so the result is trivial. (Smaller allows equal.) If A is μ-unique the set in question is a maximal subspace of X' since it is the set of f for which $\mu = 0$. The result follows from 14.0.2.

10. THEOREM. *With the notation of Remark 4, $G + X^\beta$ is closed in X'.*

By 3.0.2 the norms given in Lemmas 8, 9 are equivalent. The norm of X' is between so they are all equivalent. A variation on this proof is as follows: 4.5.2 applies with Y one-dimensional and X replaced by $G + X^\beta$, a BK space by Lemma 8. This avoids using Lemma 9.

11. COROLLARY. *With the notation of Remark 4, $G + X^\beta = c\ell X^\beta$ (in X').*

By Theorems 5 and 10.

14.2. THE FUNCTIONAL μ.

For reference we rewrite Theorem 4.4.3.

1. THEOREM. *Let A be a matrix, $X = c_A$. Then $f \in X'$ iff $f(x) = \mu \lim_A x + t(Ax) + \alpha x$, $\alpha \in X^\beta$.*

If A is μ-unique it is reasonable to write $\mu(f)$ thus defining a linear functional μ on X'. Otherwise f has representations with arbitrary μ, in particular 0. The set G was defined in 14.1.1.

2. DEFINITION. *If A is μ-unique μ is the linear functional on X' defined in Theorem 1. If A is not μ-unique, take $\mu(f) = 0$ for all $f \in X$.*

3. LEMMA. *$\mu^{\perp} = G + X^{\beta}$; A is μ-unique iff $\lim_A \notin G + X^{\beta}$ and iff $G + X^{\beta} \neq X'$.*

This is clear from Definition 2 and Theorem 1. Note that A is μ-unique iff $\mu(\lim_A) = 1$.

As stated at the beginning of this chapter, the continuity of μ in general is an open question. In this and the next section several sufficient conditions will be given.

4. THEOREM (W.H. Ruckle). *Every matrix A such that $X = c_A$ is a BK space is μ-continuous i.e. μ is continuous on X'.*

By Lemma 3 and 14.1.10 $\mu^{\perp}$ is closed, and the result follows by 5.0.1.

5. THEOREM. *Let A be a matrix such that c_A is a BK space. Then $\mu^{\perp}$ and μ-uniqueness are invariant for A.*

This means that if $c_M = c_A$ then $\mu_M(f) = 0$ iff $\mu_A(f) = 0$, and M is μ-unique iff A is. The expression for $\mu^{\perp}$ in Lemma 3 is given an invariant formulation (i.e. named) in 14.1.11. Also A is μ-unique iff $\mu^{\perp} \neq c_A'$ as in Lemma 3.

14.3. THE STRONG TOPOLOGY

In order to extend the results of the preceding section to more general matrices we need to topologize c_A'. Let X be any FK space; then *X' will be assumed to have the strong topology* (14.0.4). This is the norm topology if X is a BK space). *μ-continuity will be interpreted accordingly.*

The proof of the invariance of P (15.4.12) does not use the results of this section.

1. THEOREM. *Let* A *be a matrix,* $X = c_A$. *Then the closure of* X^β *includes* G.

Compare 14.1.5. Let E be a bounded set in X. Then $\|Ax\|_\infty \le M$ for $x \in E$ (8.0.4). Let $f(x) = t(Ax)$, choose m so that $\sum_{n=m+1}^{\infty} |t_n| < 1/M$ and let $u_k = \sum_{n=1}^{m} t_n a_{nk}$. Obviously $u \in X^\beta$. For $x \in E$,

$$|f(x)-\hat{u}(x)| = |\sum_{n=m+1}^{\infty} t_n(Ax)_n| \le 1$$

so $\hat{u} - f \in E^o$ (14.0.4) i.e. $\hat{u} \in f + E^o$. Since E^o is a basic strong neighborhood of 0 the result follows.

2. THEOREM. *These are equivalent conditions on a matrix* A, $X = c_A$: *(i)* A *is* μ-*continuous, (ii)* $\mu^\perp = c\ell\, X^\beta$, *(iii)* $G + X^\beta$ *is closed in* X', *(iv)* $G + X^\beta = c\ell\, X^\beta$.

That (iv) implies (ii), and (i) implies (iii) are by 14.2.3. That (ii) implies (i) is by 5.0.1. Finally (iii) implies (iv) by Theorem 1.

A remark of very little significance is that μ *is weak* continuous iff* $\mu = 0$, for if $\mu(f) = f(z)$ for all f then $z_k = P_k(z) = \mu(P_k) = 0$ so $z = 0$. Thus if A is μ-unique, μ cannot be weak* continuous.

3. LEMMA. *Let* X *be an FK space* $\supset bv$. *Then* χ *is continuous on* X'.

See 3.2.1 for χ. Let $E = \{1-1^{(n)}\}$. This is a bounded set in X since $1 \in B$ (10.2.7). Now if $f \in E^o$ (14.0.4), $|\chi(f)| = |\lim f(1-1^{(n)})| \le 1$ so χ is bounded on a strong neighborhood of 0, hence is continuous (14.0.3).

4. THEOREM. *Every coregular matrix* A *is* μ*-continuous.*

Let $X = c_A$. This satisfies the hypothesis of Lemma 3 by 9.6.1 and $\chi(f) = \mu(f)\chi(A)$ (9.6.8). The result is immediate from Lemma 3.

5. THEOREM. *The condition "*μ*-unique and* μ*-continuous" is invariant. Indeed* A *has this property iff* X^β *is not dense in* X', $X = c_A$.

Sufficiency: Using Theorem 1 and 14.2.3, $X' \neq \overline{X^\beta} \supset X^\beta + G = \mu^\perp$ so $\mu^\perp$ is a closed proper subspace of X'. Necessity: by Theorem 2.

6. THEOREM (J. Boos). *An almost coregular matrix is* μ*-continuous. Indeed it is sufficient that* $W \neq B$.

If it is not μ-continuous $\lim_A \in \overline{X^\beta}$ by Theorem 5. Let $x \in B$, $E = \{x\} \cup \{x^{(n)}\}$, a bounded set in X. Let $\varepsilon > 0$. There exists $f \in X^\beta$ such that $f \in \lim_A + \varepsilon E^o$. Then $|\lim_A x - \sum_{k=1}^{n} a_k x_k|$ $\leq |\lim_A x - f(x)| + |f(x) - f(x^{(n)})| + |f(x^{(n)}) - \lim_A x^{(n)}| \leq 2\varepsilon$ $+ |f(x - x^{(n)})|$. The last term $\to 0$ as $n \to \infty$ and so $\Lambda(x) = 0$. Thus $x \in W$. (13.2.5, 13.2.14).

7. COROLLARY (J. Boos). *An* AB *matrix must be* μ*-continuous.*

If not, $W = c_A$ by Theorem 6, hence $\Lambda = 0$ (13.2.14) i.e. $\lim_A x = \sum a_k x_k$ so A is not μ-unique. Thus $\mu = 0$ which is continuous. This is a contradiction.

The proof of Theorem 6 shows also that A is μ-unique; this was given earlier (13.2.25). In contrast, an AB matrix need not be μ-unique (4.4.4).

14.4. μ SPACES

The FK program is applicable to some of the results in the last section.

1. DEFINITION. *An FK space* X *is called a* μ *space if* X^β *is not dense in* X'.

By 14.3.5, a *matrix* A *is* μ*-unique and* μ*-continuous iff* c_A *is a* μ *space.* If the main conjecture is true (that every matrix is μ-continuous) then this condition characterizes μ-uniqueness as it does in any case where A is known to be μ-continuous.

2. THEOREM. *Every coregular space is a* μ *space (compare 14.3.4).*

If $u \in X^\beta$ it is immediate that $\chi(\hat{u}) = 0$. (See 14.1.1 for $\hat{u}$.) The result follows from 14.3.3 and 9.3.10.

3. THEOREM. *Let* X *be an* FK *space* $\supset \phi$. *If* $B \not\subset \overline{\phi}$ *then* X *is a* μ *space.*

Let $h \in X'$, $h = 0$ on ϕ, $h(z) = 2$ for some $z \in B$; using the Hahn-Banach theorem (3.0.1). Let $y_n = z - z^{(n)}$ and $E = \{y_n\}$, a bounded set in X. Then for any $u \in X^\beta$,

$$|h(y_n)-\hat{u}(y_n)| = |h(z) - \sum_{n+1}^{\infty} u_k z_k| \to 2$$

so $h - \hat{u} \notin E^o$ and X^β is not dense.

For $X = c_A$ this folllows from 14.3.6 and 13.2.25 since $W \subset \overline{\phi}$. The FK program for 14.3.7 is: *An FK space with AB either is a* μ *space or has AK.* This is by Theorem 3 if X does not have AD; otherwise by 10.3.19. For c_A the two cases are μ-unique and non-μ-unique.

4. EXAMPLE. *An FK space of the form* $X = c_A^o$ *cannot be a*

μ *space.* (In this assertion c_o can be replaced by any AK space.) By 4.4.2 each $f \in X'$ has the form $f(x) = t(Ax) + \alpha x$ i.e. $X' = G + X^\beta$ in the language of 14.1.2. Examination of the proof of 14.3.1 shows that $u \in \omega_A^\beta$ so, in particular, $u \in X^\beta$; where, now, $X = c_A^o$. Thus X^β is dense in X'.

5. THEOREM. *If an FK space Y has a closed subspace X which is a μ space, then Y is a μ space.*

Define $H: Y' \to X'$ by $H(f) = f|X$. Then H is onto by the Hahn-Banach theorem (3.0.1), continuous (14.0.5), and takes Y^β into X^β. Such a map preserves dense sets.

6. This is as far as we can go. Example 4 shows how to get a non-μ space which is a closed, even maximal, subspace of a μ space. The smaller space can be made conservative by making A multiplicative 0. A μ-unique multiplicative 0 triangle is shown in 13.5.1, also 13.5.2. The smaller space can also be given the form c_D by the simple device of defining D by $(Dx)_n = (Ax)_{n/2}$ if n is even, 0 if n is odd. Then $c_D = c_A^o$. In the other direction, every space is included in the non-μ space ω.

14.5. AB AND CLOSURE OF X^β

We can now give a satisfactory discussion of some results which were given in piecemeal fashion and restricted to BK spaces. The improvement of mapping and continuity of μ play an important role. We need a preliminary result which will allow improvement of mapping as in 8.3.8 to be applied when the range is not an FK space e.g. $A = 0$ in the next result. (4.0.5).

1. LEMMA. *Let A be a matrix, $X = c_A$. Then $A^T: \ell \to X^f$ is continuous.*

Since $A \in (X:c)$ it follows that $A^T \in (\ell:X^f)$ by 8.3.8. As in the case of BK spaces, X^f is given the quotient topology (14.0.6) by q where $q(f) = \{f(\delta^k)\}$ for $f \in X'$, the latter space having the strong topology (14.0.4). Let $A': c' \to X'$ be the dual map (14.0.5, A' is continuous with the strong topologies.) Note that $\ell \subset c'$ (properly) with the embedding $t \to \hat{t}$ with $\hat{t}(y) = ty$, (14.1.1, 14.1.2), and the strong topology agrees with the norm topology on ℓ (14.0.4). Now let $H = q \circ A': \ell \to X^f$. Then H is continuous as the composition of two continuous functions. The proof is concluded by showing that $H = A^T$. Let $t \in \ell$. Then $H(t)_k = q\,[A'(\hat{t})]_k = A'(\hat{t})(\delta^k) = \hat{t}(A\delta^k) = (tA)_k = (A^T t)_k$.

2. REMARK. *I conjecture that every matrix map:* $\ell \to X^f$ *is continuous if* X *is an* FK *space.* This is true if X is a BK space (14.2.8, 7.2.14). It is also true if X is reflexive or more generally if X^f has the quotient topology by X' with its Mackey topology, since then X', hence X^f, is fully complete and Ptak's closed graph theorem applies. See [80], Example 12-3-8, Theorems 12-4-5, 12-5-7.

3. THEOREM. *These are equivalent for a matrix* A *with convergent columns,* $X = c_A$. *(i)* A *has* AB, *(ii)* X^β *is closed in* X', *(iii)* X^β *is closed in* X^f, *(iv)* $X^\beta = X^f$, *(v)* $X^\beta + \phi^\perp$ *is closed in* X', *(vi)* $A^T \in (\ell:X^\beta)$.

(i) implies (ii) (Given in 10.3.11 for BK spaces): $X^\beta \supset G$ (12.5.8) hence $X^\beta = X^\beta + G = \mu^\perp$ (14.2.3). This is closed by 14.3.7.

(ii) implies (i). $X^\beta \supset G$ by 14.3.1 and the hypothesis. This implies AB (12.5.8).

(i), (iv), (vi) are equivalent. (10.4.4, 12.5.8, 13.3.1).

(iii) = (v) as in 10.3.10, using 14.0.6.

(iii) implies (vi): The rows of A, hence the columns of A^T, are in X^β and so $A^T[\phi] \subset X^\beta$. Since ℓ has AD, Lemma 1 implies the result.

(iv) implies (iii).

4. One application of Theorem 3 is to give easy examples in which X^β is not closed e.g. 13.4.10 or any non-perfect regular triangle (13.1.7). Ad hoc constructions were given in 10.1.2, 10.1.3. These are also easy examples for $X^\beta \neq X^f$. Compare 7.2.8.

5. The FK program fails for Theorem 3, for example bv^β is a proper closed subspace of bv^f. (7.3.5). This shows again that $c_A \neq bv$ as in 13.3.2.

6. All the conditions of Theorem 3 are meaningless, if A fails to have convergent columns, except (ii). It is interesting to ask what conditions equivalent to (ii) could be found. For example, suppose that $X = c_A$ has the property that $x \in X$ implies $x^{(n)} \in X$ for each n; would (i) and (ii) be equivalent in this case?

7. EXAMPLE. It was proved in 8.3.8 that $A \in (X:Y)$ implies $A^T \in (Y^\beta:X^f)$. But *A^T need not be in $(Y^\beta:X^\gamma)$ even for $Y = c$, $X = c_A$ a BK space.* Let A be any non-AB conservative triangle (e.g. 13.4.10). Then $A \in (X:c)$; if $A^T \in (\ell:X^\gamma)$ then $A^T \in (\ell:X^\beta)$ by the following argument: the columns of A^T, being the rows of A, are in X^β; X^β is closed in X^γ (4.3.18) so we may apply 8.3.6. It follows that A has AB (Theorem 3) and this is a contradiction.

8. A useful extension of the ideas mentioned in Example 7 is: *Let X be a BK space $\supset \phi$, then $A \in (X:Y)$ implies*

$A \in (F^+(X):Y^{\beta\gamma}) = (X^{f\beta}:Y^{\beta\gamma})$. We note first that if Y is a BK space this is immediate since $Y^{\beta\gamma} = Y^{\beta f}$ (10.3.8, 10.3.12, 10.1.5) and two applications of 8.3.8 give the result. In the general case (Y any set of sequences) let $u \in Y^\beta$, $h_n(x) = \sum_{j=1}^{n} u_j(Ax)_j$ for $x \in X$. By hypothesis this converges as $n \to \infty$ so by uniform boundedness (1.0.3) $|h_n(x)| \leq K\|x\|$. Now fix $x \in F^+$ and apply this inequality to $x^{(r)}$. Since x has bounded sections, $C \geq K\|x^{(r)}\| \geq |h_n(x^{(r)})| = |\sum_{k=1}^{r} x_k \sum_{j=1}^{n} u_j a_{jk}|$. We may allow $r \to \infty$ since $x \in \omega_A$ by 12.2.3; this says $C \geq |h_n(x)|$ for all n and so $Ax \in Y^{\beta\gamma}$.

9. EXAMPLE. A special case of Remark 8 is $(c_0:Y) \subset (\ell^\infty:Y^{\beta\gamma})$.

CHAPTER 15
THE SUBSPACE P

15.0. FUNCTIONAL ANALYSIS

1. The sum of a closed and a finite dimensional subspace is closed. Thus if E is a subspace the closure of $E \oplus u$ is $\overline{E}$ or $\overline{E} \oplus u$ [80] Theorem 6-3-3.

2. $Y^{\#}$ denotes the *algebraic dual* of a vector space Y i.e. the set of all linear functionals. Let Z be a total subspace of $Y^{\#}$. For $H \subset Z$ let $H^{\perp} = \{y \in Y: h(y) = 0$ for all $h \in H\}$; for $E \subset Y$ let $E^{\perp} = \{g \in Z: g(y) = 0$ for all $y \in E\}$. Then $H^{\perp\perp}$ is the weak $* = \sigma(Z,Y)$ linear closure of H [80] #8-3-110.

3. Each $g \in Y'^{\#}$ which is weak * continuous is given by $y \in Y$ i.e. $g(f) = f(y)$ for all $f \in Y'$ [80] Theorem 8-1-7.

4. Let X be a Banach space with separable dual and S a subspace of X. Then, in the weak * topology of X'', each point in $\overline{S}$ is a sequential limit point of S [81], Lemma 3.3.

5. Let f, g be linear functionals such that $g(x) = 0$ implies $f(x) = 0$. Then f is a multiple of g. [80] Theorem 1-5-1.

15.1. T AND THE TEST FUNCTIONS

The subspace P of c_A (15.2.1) was introduced by H. R. Coomes and V. F. Cowling to generalize the Type M condition to non-triangles. (See 15.2.2.) The question of its invariance (15.4.12) was settled affirmatively in [6]. A simpler proof (15.2.10) is available in the presence of μ-continuity which, as far as we know, covers all cases.

We begin by associating with a matrix a certain subset of ℓ:

1. DEFINITION. *Let* A *be a matrix. Then*

$$T = T_A = \{t \in \ell\colon (tA)x \text{ exists for all } x \in X = c_A\} = \{t \in \ell\colon tA \in X^\beta\}.$$

Clearly $T = \ell$ *iff* $L_e = X$ *(12.2.2), and this* (if A has convergent columns) *is equivalent to* AB (13.3.1). One thinks of Type M in connection with Definition 1 since if $t \perp A$, certainly $t \in T$.

2. DEFINITION. *A function* $f \in X'$, *where* $X = c_A$, *is called a test function for* A *if* $\mu(f) = 0$ *and* $f = 0$ *on* ϕ. (It is understood that $X \supset \phi$).

3. EXAMPLE. *If* A *is not replaceable or not* μ*-unique then every function vanishing on* ϕ *is a test function.* In the first case $\mu(f) = 0$ by 13.2.18. In the second case it is by definition.

4. EXAMPLE. *Let* A *be almost coregular or, more generally, satisfy* $W \neq B$. *Then every function vanishing on* B *is a test function.* If not, let $f = 0$ on B, $\mu(f) = 1$. For $x \in B$, $f(x) = \lim_A x = \gamma x$ (12.5.9) and so $\lim_A x = -\gamma x$ for $x \in B$. Taking $x = \delta^k$ gives $\gamma_k = -a_k$ and so $\lim_A x = ax$ for $x \in B$. This implies $W = B$ by 13.2.5, 13.2.14.

The connection between the two definitions follows:

5. THEOREM. *Let* A *be a matrix with convergent columns. Then* f *is a* **test** *function for* A *iff* $f(x) = t(Ax) - (tA)x$ *for some* $t \in T$.

Sufficiency is trivial. Now suppose that f is a test function. Then $f(x) = t(Ax) + \alpha x$ (4.4.3). Also $0 = f(\delta^k) = (tA)_k + \alpha_k$ and so $tA \in X^\beta$ since α does. Hence $t \in T$ and the representation of f follows.

We shall see that the set of test functions is invariant. (15.4.11). The proof is complicated and it is worthwhile to prove a (possibly) special case:

6. THEOREM. *Let A be a matrix with convergent columns such that all equipotent matrices are μ-continuous. Then the set of test functions is invariant for A.*

By 14.3.2, $\mu^{\perp}$ is invariant; trivially ϕ is invariant.

Paradoxically, in view of Theorems 5 and 6, T is not invariant, even for equipotent triangles in which case μ-continuity is assured by 14.2.4. ([46], p. 240.)

Many sufficient conditions for μ-continuity are given in §14.3.

15.2. THE SUBSPACE P.

1. DEFINITION. *Let A be a matrix. Then $P = P_A = \{x \in c_A: (tA)x = t(Ax)$ for all $t \in T\}$ (15.1.1).*

2. The original motivation for the definition and name of P was Theorem 6 which is the analogue of the Type M condition for general matrices. It was originally given for conservative matrices.

3. THEOREM. *Let $X = c_A \supset bv$ (e.g. if A is vsc). Then P is closed and includes the perfect part of X and all the other distinguished subspaces (hence their closures also).*

For each $t \in T$ set $f_t(x) = (tA)x - (tA)x$. This is continuous by the Banach-Steinhaus theorem (1.0.4) and $P = \cap\{f_t^{\perp}: t \in T\}$. Next $B = L_e$ (12.5.3) $\subset P$ trivially. Also $\phi_1 \subset B$ by 10.2.6, 10.2.7 and the rest is trivial.

Possibly P is strictly larger than $\overline{B}$ and the perfect part.

See Examples 5 and 9. Note also that *the definition of* P *covers all matrices* — not just those with convergent columns as for the other subspaces.

4. THEOREM. *Let* A *be a matrix with convergent columns. Then* $P = \cap\{f^{\perp}: f$ *is a test function for* $A\}$.

From 15.1.5.

5. EXAMPLE. *Let* A *be almost coregular or more generally satisfy* $W \neq B$. *Then* $P = \overline{B}$. *If* A *is coregular,* P *is the perfect part of* c_A. It follows from 15.1.4, Theorem 4 and the Hahn-Banach theorem (3.0.1) that $P \subset \overline{B}$. The rest is by Theorem 3 and 13.1.5.

6. THEOREM. *A coregular matrix* A *is perfect iff* $P = c_A$.

From Example 5.

7. EXAMPLE. *Let* A *be a matrix with convergent columns. If* A *is not replaceable or not* μ*-unique then* $P = \overline{\phi}$ by 15.1.3 and the Hahn-Banach theorem (3.0.1). *If* A *is not* μ*-unique then* $\overline{\phi} = W = F = B = P$ by 13.2.25.

8. THEOREM (Cathy Madden). *Suppose that a matrix* A *has a right inverse* A' *with columns in* P. *Then* A *is of Type M.*

Let $t \perp A$. Then $t \in T$ and so for $x \in P$ we have $t(Ax) = (tA)x = 0$. Taking x to be any one of the columns of A' yields $t = 0$.

A crude form of this result was given in 3.3.9. If A is conservative $B^+ \supset B^+(c)$ (10.2.9) $= \ell^{\infty}$, so any bounded sequence is in P (Theorem 3). (The more general assumption $A \in \Phi$ of 3.3.8 is covered by using ℓ^{∞}_A.)

Thus a P-perfect matrix (i.e. $P = c_A$) with a right inverse must be of Type M.

9. EXAMPLE. *A conull triangle A with* $P \neq \overline{B}$ = *perfect part of* c_A. Let A be the matrix of 13.5.2 with $W = F = B = X \cap \ell^\infty \subset$ perfect part of X, where $X = c_A$. With $f = \lim_A$ we have $f = 0$ on B but $f(u) = 1$ where $u = \{n\}$. Thus $\overline{B} \neq X$. Since the perfect part $\supset B \supset \phi_1$ *the perfect part is* $\overline{B}$. The result follows when it is proved that $P = X$.

Let $t \in T$, $x \in X$. We have to prove that $(tA)x = t(Ax)$. Now using Abel's identity (1.2.9), $\sum_{k=1}^{n} (tA)_k x_k = \sum (t_k - t_{k+1}) x_k$ $= \sum_{k=1}^{n} t_k(x_k - x_{k-1}) - t_{n+1}x_n = U_n - V_n$, say. Since $U_n \to t(Ax)$ we know that $V = \lim V_n$ exists and we have to prove that it is 0. Taking $x = \{n\}$ yields that $\lim n\, t_{n+1}$ exists and since $t \in \ell$ this implies that

$$\lim n\, t_{n+1} = 0 \tag{1}$$

Now let $y = Ax$ so that $y \in c$. Then $V_n = t_{n+1} \sum_{k=1}^{n} y_k = nt_{n+1}C_n(y)$ where C is the (C,1) matrix. This tends to 0 by (1) since C is regular. *This matrix is* μ-*unique* by Example 7.

The matrix in Example 9 is a special case $(r_n = n)$ of A (6.3.2). R. DeVos has calculated that the result of Example 9 holds for each such A.

10. THEOREM. *Let A be a matrix with convergent columns such that all equipotent matrices are* μ-*continuous. Then P is invariant for A.*

This is immediate from Theorem 4 and 15.1.6.

11. EXAMPLE. *Let A be a reversible matrix with null columns and* $P = c_A$. *Then A is* μ-*unique.* (The assumption of

null columns cannot be omitted, 4.4.4.) If not $\lim_A = 0$ on ϕ, hence on P (Example 7). Thus $x \notin P$ if $Ax = 1$.

A converse for Example 11 holds:

12. EXAMPLE (R. DeVos). *Let A be a reversible matrix with null columns and $\overline{\phi} \supset c_A^o$, equivalently A is of Type M. Then A is μ-unique iff $P = c_A$.* Half is Example 11. Conversely suppose that $z \notin P$ i.e. $t(Az) \neq (tA)z$ for some $t \in T$. Let $f(x) = t(Ax) - (tA)x$. Now $P \supset \overline{\phi}$ (Theorem 4) so $f = 0$ on c_A^o, hence f is a non-zero multiple of $\lim_A$ (15.0.5); non-zero since $f(z) \neq 0$. Thus A is not μ-unique.

13. This suggests an extremely tantalizing question: *must a multiplicative triangle be μ-unique?*

14. Consider the non-μ-unique triangle A of 4.4.4. Here $X = c_A = a^\beta$ has AK so A has AB, hence $P = X$. An equipotent multiplicative matrix M is shown in 5.2.7. Example 11 shows paradoxically that *no such M can be reversible*, even though $c_M = X$ is a BK space, since all the conditions (except reversible) are invariant. (Theorem 10 and 14.2.5).

On the other hand, *a μ-unique replaceable triangle has an equipotent multiplicative triangle.* (15.5.4)

What has been proved so far in the direction of the invariance of P? Sufficient conditions are: c_A is a BK space (Theorem 4, 15.1.6, 14.2.4. This almost names P); A is coregular (Example 5; $W \neq B$ is sufficient and P is named); A is not replaceable (Example 7; P is named); A has AB (Theorem 3; $P = c_A$). However Example 7 does not yield invariance of P for non-μ-unique matrices since this condition has not yet been proved invariant (15.4.9).

15.3. FACTORIZATION

If A, M are positive integers with $M > A$, $M = EA + D$ with $0 \leq D < A$. Without the last restriction we could take $E = 0$, $D = M$. If A, M are matrices with $c_M \supset c_A$ it is often possible to write $M = EA + D$ where D is "small" (or c_D is "large"). For example if A, M are triangles we may take $E = MA^{-1}$, $D = 0$. In general it is not to be expected that $D = 0$, for example if the first column of A is 0. Such results are called *factorization* or *quotient* theorems. Some valuable ones are given in [3] (D has small norm, is coercive, etc.) and [58], ($D = 0$).

Now if m, e, d represent the nth row of M, E, D, respectively, the equation $M = EA + D$ is the same as $m = eA + d$ for each n. This will follow from $mx = e(Ax) + dx$ by setting $x = \delta^k$. Thus the problem is to represent the row function mx (11.1.1) which is defined on c_A, and this is easy. In fact we could take $e = 0$, $d = m$ but to obtain the "smallness" of D will require a little more subtlety. (The next two results may be omitted.)

1. LEMMA. *Let A be a μ-unique matrix, $X = c_A$, $m \in X^\beta$, $h(x) = mx$. Then for $x \in X$, $h(x) = e(Ax) + dx$, $e \in \ell$, $d \in \omega_A^\beta$.*

This is immediate from 4.4.3 and 4.4.4. That $\mu = 0$ is because h has the representation mx.

2. COROLLARY. *Let A be a μ-unique matrix with convergent columns. Let $c_M \supset c_A$. Then $M = EA + D$ with the rows of E in ℓ and the rows of D in ω_A^β.*

Apply Lemma 1 to each row of M and set $x = \delta^k$.

This result is given for comparison with the next one which is tailor-made for its application to proving invariance theorems.

One can see that Corollary 2 is not strong enough by considering A to be a non-AB matrix and taking $M = A$. Then $E = 0$, $D = A$ will satisfy Corollary 2 but not Theorem 3. (Take $Y = c$ in 12.5.3).

3. THEOREM. *Let A be a μ-unique matrix with convergent columns, $Z = c_A$. Let $c_M \supset Z$. Then $M = EA + D$ where $\|E\| < \infty$, $(tD)x = t(Dx)$ for $t \in \ell$, $x \in \omega_A$.*

Thus the smallness condition on D is that every $x \in \omega_A$ has the associativity property.

Let $h_n(x) = (Mx)_n$. Then $\{h_n\}$ is pointwise convergent on Z hence equicontinuous (7.0.2). In the construction of 4.4.1 with $f = h_n$, $X = \omega_A$, $Y = c$, we have $|h_n(x)| \leq K \sum_{i=1}^{r} p_i(x) + L\|x\|_A$ with K, L, r independent of n and $x \in Z$. Here $\{p_n\}$ are *all* the seminorms of ω_A (listed as $\{p_n, h_n\}$ in 4.3.13.) Hence $h_n = F_n + g_n \circ A$ with $F_n \in \omega_A'$, $|F_n(x)| \leq K \sum p_i(x)$; $g_n \in c'$, $|g_n(y)| \leq L\|y\|_\infty$ for $y \in c$ as in 4.4.1.

Since ω_A has AK (4.3.8), $F_n(x) = \sum F_n(\delta^k)x_k = (Dx)_n$ where $d_{nk} = F_n(\delta^k)$. Also $g_n(y) = \mu_n \lim y + \sum e_{nk}y_k$ for $y \in c$ (1.0.2) so $h_n(x) = \mu_n \lim_A x + e_n(Ax) + d_n x$ where e_n, d_n are the nth row of $E = (e_{nk})$ and D, respectively. Since also $h_n(x) = (Mx)_n$ and A is μ-unique it follows that $\mu_n = 0$ and so finally $Mx = E(Ax) + Dx$. Taking $x = \delta^k$ shows that $M = EA + D$. That $\|E\| < \infty$ is because $L \geq \|g_n\| = \sum |e_{nk}|$. Turning to D, let $t \in \ell$ and set $u(x) = t(Dx)$ for $x \in \omega_A$; this quantity exists since $\sum |t_n(Dx)_n| = \sum |t_n F_n(x)| \leq K \sum p_i(x) \cdot \|t\|_1$, hence $u \in \omega_A'$ by the Banach-Steinhaus theorem (1.0.4). Since ω_A has AK (4.3.8), $u(x) = \sum u(\delta^k)x_k = (tD)x$.

The condition on D is related to bounded sections by 12.5.3 with $Y = \ell^\infty$; it is easy to see that D has bounded columns so that this makes sense.

4. THEOREM. *The matrices in Theorem 3 have the properties* $(EA)x = E(Ax)$ *for* $x \in Z$ *and* $(tE)y = t(Ey)$ *for* $t \in \ell$, $y \in c$.

The first condition holds because in the course of proving Theorem 3 the identities $Mx = E(AX) + Dx$ and $M = EA + D$ were proved. The latter gives $Mx = (EA)x + Dx$ and the result follows.

The second condition holds because $\ell^\infty_E \supset \ell^\infty$ and so $B[\ell^\infty_E] \supset B[\ell^\infty] = \ell^\infty \supset c$, using 10.2.9. The rest is by 12.5.3 with $Y = \ell^\infty$.

Sources: [6], [18], [81].

15.4. INCLUSION AND INVARIANCE

The factorization theorem just proved will allow us to complete the discussion of μ, P and the test functions in a satisfactory way.

1. DEFINITION. *Let* A, M *be matrices with* $c_M \supset c_A$. *Then* $\mu_A(M)$ *is* $\mu(\lim_M|c_A)$; M *is called coregular (mod A) if* $\mu_A(M) \neq 0$, *otherwise conull (mod A).*

2. EXAMPLE. Take $A = I$. Then $\lim_M x = \chi(M)\lim x + tx$ for $x \in c_A = c$ (1.3.8). Thus $\mu_I(M) = \chi(M)$ *whenever* M *is conservative and so coregular (mod I) = coregular.*

3. EXAMPLE. Let A be coregular. Then $\chi(M) = \mu_A(M)\chi(A)$ (9.6.8) and so M is coregular iff it is coregular (mod A).

4. A *is* μ*-unique iff* $\mu_A(A) \neq 0$ *i.e. iff* A *is coregular (mod A). If* A *is not* μ*-unique and* $c_M \supset c_A$ *then* M *is conull (mod A), in particular* A *is conull (mod A).*

5. REMARK. In the remainder of this section A *is a matrix with convergent columns.* We shall eschew the usual reference to this remark — caveat lector.

6. THEOREM (W. Beekmann). *Let* $c_M \supset c_A$ *and* $f \in c'_M$ *so that* $f(x) = \mu\, lim_M x + t(Mx) + \alpha x$. *(4.4.3). Then* $\mu(f|c_A) = \mu\cdot\mu_A(M)$.

If A is not μ-unique both sides are 0. Now assume that A is μ-unique. Fix some particular value of μ (so that it is irrelevant whether M is μ-unique.) For $x \in c_A$ we have $\lim_M x = \mu_A(M)\lim_A x + u(Ax) + vx$, $u \in \ell$, $v \in c_A^\beta$ (4.4.3) and so for $f \in c_A$

$f(x) = \mu\cdot\mu_A(M)\lim_A x + \mu u(Ax) + \mu vx + t(Mx) + \alpha x = \mu\cdot\mu_A(M)\lim_A x + g(x) + h(x)$, where $g(x) = t(Mx)$, $h(x) = \mu u(Ax) + (\mu v+\alpha)x$. Now $\mu(h) = 0$ since $\mu u \in \ell$ and $\alpha \in c_M^\beta \subset c_A^\beta$ so that $\mu v + \alpha \in c_A^\beta$. The result will follow when it is shown that $\mu(g) = 0$. Now, applying 15.3.3 and 15.3.4 we have $g(x) = t[(EA+D)x] = t[(EA)x] + t(Dx) = (tE)(Ax) + (tD)x$. This is a representation of $g \in c'_A$ with $\mu = 0$, for $tD \in c_A^\beta$ (since $(tD)x$ exists) and $tE \in \ell$ (since $\|E\| < \infty$ — by direct calculation or by 8.3.1(d) with $A = E^T$).

The essence of Theorem 6 may be summarized in the formula $G[M] \subset G[A] + c_A^\beta$, where G is given in 14.1.1.

7. COROLLARY. *If* A *is* μ-*unique* (μ-*continuous*), *then every matrix* M *which is coregular(mod A) is* μ-*unique* (μ-*continuous*).

The theorem is true (but vacuous by Remark 4) if A is not μ-unique. Taking $f = \lim_M$ in Theorem 6 gives $\mu_A(M) = \mu\cdot\mu_A(M)$ where $\mu = \mu_M(M)$. Thus $\mu = 1$ and M is μ-unique. Now suppose that A is μ-continuous and, with no loss of generality, that $\mu_A(M) = 1$. By Theorem 6, $\mu_M(f) = \mu_A(f|c_A)$; now if $f \to 0$ in the strong topology on c'_M it follows that $f|c_A \to 0$ in the strong topology on c'_A (14.0.5) and so $\mu_M(f) \to 0$.

This generalizes the earlier result that a coregular matrix

must be μ-unique and μ-continuous (14.4.2 which, however, holds in the FK setting). Of course a μ-unique and continuous conservative matrix need not be coregular(mod I) (13.5.1, 13.5.6).

From these results it follows that various properties behave properly with respect to inclusion; in particular, are invariant:

8. THEOREM. *Let* $c_N \supset c_M \supset c_A$. *If* M *is conull(mod A) so is* N. *If* M *is coregular(mod A) and* c_M *is closed in* c_N, *then* N *is coregular(mod A). In particular, coregular(mod A) and conull(mod A) are invariant.*

Compare 9.3.6. For the first part take $f = \lim_N | c_M$ in Theorem 6 which gives $\mu_A(N) = \mu_M(N)\cdot\mu_A(M)$. For the second part let $g = \lim_M$ and extend g to $f \in c_N'$ by the Hahn-Banach theorem (noting that g is continuous on its domain by 4.2.5). Now apply Theorem 6 to f, replacing M by N. This yields $\mu_A(M) = \mu \cdot \mu_A(N)$ and so neither number on the right can be 0.

9. COROLLARY. *μ-uniqueness is invariant.*

Let A be μ-unique and $c_M = c_A$. Then A is coregular (mod A) (Example 4), hence M is coregular(mod A) (Theorem 8) and consequently μ-unique. (Corollary 7).

10. COROLLARY. *μ-continuity is invariant.*

Like Corollary 9 or by Corollary 9 and 14.3.5. Note that any non-μ-unique matrix is μ-continuous.

In some earlier results the hypotheses were that every equipotent matrix is μ-continuous (15.1.6, 15.2.10). By Corollary 10 it is sufficient to assume that the matrix in question is μ-continuous. However, the conclusions of these theorems will now be obtained without even this assumption:

11. THEOREM. *Let* $c_M \supset c_A$. *Then each test function* f *for* M *is a test function for* A *i.e.* $f|c_A$ *is a test function for* A. *In particular the set of test functions is invariant.*

By Thoerem 6, $\mu(f|c_A) = 0$.

12. THEOREM. *If* $c_M \supset c_A$, *then* $P(M) \supset P(A)$. *In particular,* P *is invariant.*

By Theorem 11 and 15.2.4. Thus P joins the other distinguished subspaces in being monotone (10.2.9).

13. THEOREM. *For* $X = c_A$, $P = \overline{\phi}$ *or* $P = \overline{\phi} \oplus u$. *Both cases may occur even if* A *is a coregular conservative triangle.*

Let $Z = X''$ if A is μ-continuous, otherwise let Z be the algebraic dual of X' i.e. the set of all linear functionals. We may consider $X \subset Z$ by the embedding $x \to h$ where $h(f) = f(x)$. Also $\mu \in Z$. Let $P^+ \subset Z$ be the set $(\phi^\perp \cap \mu^\perp)^\perp$ i.e.

$$\{h \in Z: h(f) = 0 \text{ whenever } \mu(f) = 0 \text{ and } f = 0 \text{ on } \phi\}.$$

Thus $P^+ = H^{\perp\perp}$ where $H = \phi \oplus \mu$, and this is $\overline{\phi}$ or $\overline{\phi} \oplus \mu$ (15.0.1, 15.0.2) with the closure taken in the weak * topology of Z. Hence $\overline{\phi}$ is a maximal subspace of P^+ or is all of P^+. Since $P = P^+ \cap X$ (15.2.4) and, on X, the weak * topology of Z coincides with the weak topology of X, the first statement of the theorem follows with the closure taken in the weak topology. But this is the same, since ϕ is a subspace, as the closure in the FK topology (5.0.2). For the second part consult 15.2.7 with 5.2.5; also observe that *if* A *is regular,* $1 \notin \overline{\phi}$ (5.2.1) but $1 \in P$ (indeed $1 \in F$ by 10.2.7) hence $P = \overline{\phi} \oplus 1$.

Sources: [5], [6], [46], [81].

15.5. REPLACEABILITY

In this section we complete the "function as matrix" theorem 5.1.1 and give necessary and sufficient conditions for replaceability.

1. REMARK. In this section A is a matrix with convergent columns, $X = c_A$.

2. THEOREM. *Let* $f \in X'$ *(Remark 1). If* A *is not* μ*-unique, there exists an equipotent matrix* M *such that* $\lim_M = f$. *If* A *is* μ*-unique, such* M *exists iff* $\mu(f) \neq 0$.

The first part is by 5.1.4. Sufficiency in the second part is by 5.1.1. Necessity is because $\mu(f) = \mu_A(M) \neq 0$ by 15.4.8 and 15.4.4.

3. THEOREM (W. Beekmann, S-Y. Kuan). *With* A *as in Remark 1, if* A *is not* μ*-unique it is replaceable. If* A *is* μ*-unique it is replaceable iff there exists* $f \in X'$ *with* $f = 0$ *on* ϕ, $\mu(f) \neq 0$.

The first part is by 13.2.18. For the second part, if M is an equipotent multiplicative matrix, $f = \lim_M$ satisfies the requirements as in 15.4.8. Conversely if such f exists, A is replaceable by 5.1.1.

4. *If* A *is a* μ*-unique replaceable triangle it has an equipotent multiplicative triangle.* (Compare 15.2.14). Let f be as in Theorem 3. Then $f(x) = \mu \lim_A x + t(Ax)$ (4.4.3) with $\mu \neq 0$. The procedure of 5.1.1 yields a triangle since $\alpha = 0$ so that the proof ends with the definition of D (1.7.15.)

5. J. Boos has pointed out that *if* A *is conservative and* $A^T[\ell]$ *contains* a, *then* A *is replaceable.* Here $a_k = \lim a_{nk}$;

note that $A^T \in (\ell:\ell)$ by 8.3.8. The proof is that with $a = A^T t$, $f(x) = \lim_A x - t(Ax)$ defines f satisfying the conditions of Theorem 3. The *converse holds if* A *is* μ*-unique and either reversible, or one to one and row-finite* since with f as in Theorem 3, $f(x) = \mu \lim_A x + t(Ax)$ (5.4.3, 5.5.3) with $\mu \neq 0$. Then $0 = f(\delta^k) = \mu a_k + (A^T t)_k$. *The converse fails for a non-*μ*-unique triangle* as may be easily checked for the matrix of 4.4.4. I do not know what happens in other cases.

6. Under what circumstances is $c_A = c_M^o$ for some M? The answer is: iff A is not μ-unique. $\Rightarrow$: By Theorem 2 with $f = 0$. $\Leftarrow$: Define a matrix D such that $c_D = c_M^o$ by the device shown in 14.4.6. Now $\lim_D = 0$ so D is not μ-unique. Hence A is not (15.4.9). Note that this gives an alternate proof of 14.4.4.

7. COROLLARY. *With* A *as in Remark 1,* A *is replaceable and* μ*-unique iff* μ *does not belong to the weak * closure of* ϕ.

By Theorem 3 and 15.0.3 with the Hahn-Banach theorem (3.0.1) applied to $X'^{\#}$.

Corollary 7 is an invariant condition. Note also that by 15.4.6, any two possible μ's (for equipotent matrices) are linearly dependent.

8. COROLLARY. *With the notation of Remark 1,* A *is replaceable,* μ*-unique and* μ*-continuous iff* X^β *is not dense in* $\phi^\perp$.

Necessity: Theorem 3 says that $\mu^\perp \not\supset \phi^\perp$ and this implies the result by 14.3.2. Sufficiency: the second and third conditions hold by 14.3.5. Hence $\mu^\perp = \overline{X^\beta}$ (14.3.2) $\not\supset \phi^\perp$ and A is replaceable by Theorem 3.

9. Corollary 8 *names* the conjunction of the three properties

and so we are led to study FK spaces X with the property (*) X^β *is not dense in* $\phi^\perp$. If $X = c_A$, X has (*) iff A is replaceable, μ-unique and μ-continuous. For coregular matrices (which are automatically μ-unique and μ-continuous by 9.6.10 and 14.3.4) the condition (*) is equivalent to replaceability which, in turn is equivalent to the condition $1 \notin \overline{\phi}$ (13.2.16).

Thus there are two generalizations to FK spaces of replaceability for coregular matrices. I do not know whether these conditions are dependent for general coregular FK spaces.

10. If A is coregular and not replaceable, no stronger coregular matrix B can be replaceable. This is immediate from 5.2.1, 4.2.4. Coregular may not be omitted, for example take $B = 0$.

11. EXAMPLE. *A Mercerian triangle* T *and a coregular triangle* A *such that* ATA^{-1} *is not conservative.* Thus $c_{AT} \not\supset c_A$ (1.7.5). Compare this with $c_{TA} = c_A$ (1.7.5). Let A be non-replaceable, $\chi(A) = 1$ (5.2.5). Let M be Mazur's matrix (1.8.12) using the column limits of A as the sequence t i.e. $t_k = a_k$. Then A is consistent with M since they have the same column limits and value of χ. This implies that AM^{-1} is regular (1.7.10) and so by Remark 10 it is not stronger than A i.e. $AM^{-1}A^{-1}$ is not conservative (1.7.5). With $T = M^{-1}$ the example is concluded.

Sources: [5], [43].

15.6. MISCELLANY AND QUESTIONS.

Once again we deal only with matrices with convergent columns.

1. If A is a triangle it is very easy to compute μ in the expression $f(x) = \mu \lim_A x + t(Ax)$, (4.4.5), namely define $g \in c'$ by $g(y) = f(A^{-1}y)$ then $\mu = \chi(g) = g(1) - \sum g(\delta^k) = f(u) - \sum f(v^k)$ where u is the sequence of row-sums of A^{-1} and v^k is the kth column of A^{-1}. This fact can be used to give easy proof *for triangles* of such results as 15.4.8.

2. THEOREM (Cathy Madden). *A matrix is μ-unique iff every equipotent matrix is weakly μ-unique.*

Necessity is by invariance (15.4.9). Conversely if A is not μ-unique there is an equipotent matrix M with $\lim_M = 0$ (5.1.1) This M is not weakly μ-unique.

3. THEOREM. *A matrix A is μ-unique iff $\chi \in \sigma(c'',c')$-closure of $M[X]$ for every equipotent M; $X = c_A = c_M$.*

If A is not μ-unique the matrix M given in Theorem 2 has range $\{0\}$. Conversely we may assume $M = A$ by invariance (15.4.9). If the conclusion is false, there exists $g \in c'$ with $g = 0$ on $A[X]$ and $\chi(g) = 1$. For $y \in c$, $g(y) = \lim y + ty$. Let $f(x) = \lim_A x + t(Ax)$ for $x \in X$. Then $f(x) = g(Ax) = 0$ for all x so A is not μ-unique.

4. If A is a triangle, $A[X] = c$ which is dense in c'' (Theorem 3) so the condition of Theorem 3 for *some* M is not sufficient. In all the cases of non-μ-uniqueness we have seen, it was revealed by an equipotent matrix M with $M[X] \subset c_o$ whose closure does not contain χ.

As a corollary we can find yet another sufficient condition for μ-continuity. The condition is satisfied in two special cases (14.1.3) in which however μ-continuity is assured because the domains are BK spaces (14.2.4).

5. COROLLARY. *If* $X^\beta \subset G$, *then* A *is* μ-*continuous.* *(*$X = c_A$*).*

We may assume that A is μ-unique. By Theorem 3, there is a sequence $\{x^n\}$ in X such that $g(Ax^n) \to \chi(g)$ for all $g \in c'$ (15.0.4). For each $f \in X'$ we shall prove that $f(x^n) \to \mu(f)$. By hypothesis we have $f(x) = \mu \lim_A x + t(Ax)$. Let $g(y) = \mu \lim y + ty$ for $y \in c$. Then $f(x^n) = g(Ax^n) \to \chi(g) = \mu$. The rest is as in the proof of 14.3.3 with $E = \{x^n\}$, a (weakly) bounded set.

6. Here are some questions and suggestions for further study:

Is P invariant for a matrix whose columns are not convergent?

What is the FK equivalent for μ-uniqueness? This may be answered by 14.3.5. Matrix-bound conditions were given in 14.2.3, and Theorems 2, 3.

Solve the naming problem for μ-continuity, the set of test functions and P.

Give an FK equivalent for coregular(mod A) analogous to 4.6.2.

7. To prove μ-continuity for A it is sufficient to find a μ-continuous M such that A is coregular(mod M) (15.4.7). Any M such that c_M is a BK space will do (14.2.4). Theorem 3.5.1 looks promising in this direction; however the conditions do not imply that A is coregular(mod M). For example if $Mx = \{x_n/n\}$, $Ax = \{x_n/n^2\}$ then $c_M \cap \ell^\infty = c_A \cap \ell^\infty$ but A is conull(mod M). (Also A is μ-unique — otherwise $A = 0$ would do as well.)

8. Cathy Madden has suggested a study of what she calls R-invariance in which properties common to A, M are studied if

$c_A = c_M$, $\omega_A = \omega_M$. (The FK program would deal with the properties of pairs X, Y of FK spaces with $X \subset Y$, Y an AK space.)

To give one example: *weak* μ-*uniqueness is R-invariant.* (Compare 5.4.12)

Sources: [47], [81].

CHAPTER 16
SEQUENTIAL COMPLETENESS AND SEPARABILITY

16.0. FUNCTIONAL ANALYSIS

1. Let X be a Banach space. Then X' with its weak * topology is sequentially complete [80], Examples 9-3-8, 9-3-2.

2. The sup (4.0.9) of two pseudometrizable separable vector topologies is separable. (A pseudometric is the same as a metric except that $d(x,y) = 0$ is allowed.) [80A], #6.7.114.

3. (N.J. Kalton's closed graph theorem.) Let Y be a locally convex space with Y' weak * sequentially complete. Let X be a separable FK space. Then each linear map from Y to X which has closed graph is weakly continuous. [80], Theorem 12-5-13.

4. Let X be a vector space and Z a total subspace of $X^{\#}$ (15.0.2). For $E \subset Z$, $E^{o} = \{x \in X: |f(x)| \leq 1 \text{ for all } f \in E\}$. The set of all E^{o}, E a finite set in Z, is a local base of neighborhoods of 0 for a topology on X called $\sigma(X,Z)$. In this topology, $x \to 0$ iff $f(x) \to 0$ for all $f \in Z$. [80], Examples 8-5-4, 4-1-9; Definition 4-1-10.

5. Continuing 4: The set of all E^{o}, E an absolutely convex weak * compact set in Z is a local base of neighborhoods for a topology on X called $\tau(X,Z)$. It is the largest locally convex topology such that $X' = Z$. [80], Theorem 9-2-3.

6. With the topology $\tau(\ell^{\infty}\ \ell)$, ℓ^{∞} is an AK space. [80], #9-5-107.

7. Let X be a locally convex sequence space $\supset \phi$ with topology larger than the relative topology of ω. Then a set in X is compact iff it is sequentially compact. [30], p. 1010, Theorem 6. It follows that two such topologies with the same convergent sequences have the same compact sets.

8. Let X, Z be as in 4, and let a subspace V of Z have the property that $\sigma(X,V)$ and $\sigma(X,Z)$ have the same compact sets. Then $\tau(Z,X)|_V = \tau(V,X)$. This is clear from 5 in which now the sets E are in X, and E^o in V or Z.

9. Let X, Y be locally convex spaces and $T: X \to Y$ a weakly continuous linear map. Then T is continuous when X, Y have $\tau(X,X')$, $\tau(Y,Y')$ respectively. [80] Example 11-2-5.

10. The topology of an FK space X is $\tau(X,X')$. [80], Theorem 10-1-9, Example 10-1-10.

16.1. SEQUENTIAL COMPLETENESS

The role of sequential completeness was pointed out by G. Bennett and N.J. Kalton in [12] in which they proved Theorem 10 of this section. Their tools were the deepest methods of functional analysis — most of their results are not presented here. The proof of Theorem 10 given here is due to the author and A.K. Snyder. Motivation for the result is contained in Remark 14 and the next section.

1. REMARK. *In this section* V *is a vector space of sequences satisfying* $c_o \subset V \subset \ell^\infty$.

2. DEFINITION. *A sequence* $\{a^n\}$ *of members of* ℓ *is called V-Cauchy if* $\lim f(a^n)$ *exists for each* $f \in V$*, where* $f(a)$

means $\sum f_k a_k$.

3. LEMMA. *A sequence* $\{a^n\}$ *in* ℓ *is* *V-Cauchy iff* $c_A \supset V$ *where* $A = (a_k^n)$.

For $x \in V$, $x \in c_A$ iff $\lim \sum_k x_k a_k^n$ exists i.e. (Definition 2) iff $\lim x(a^n)$ exists.

4. Note that a^n is just the nth row of A. Since $V \supset c_o$, such A has convergent columns and $\|A\| < \infty$. In particular $\{a^n\}$ is norm bounded in ℓ, $a_k = \lim a_k^n$ exists for each k and $a \in \ell$ (1.3.7).

5. EXAMPLE. Let $V = c$. Then $\{\delta^n\}$ is V-Cauchy since $A = I$ (Lemma 3). Here $a = 0$ (Remark 4) and it is false that $f(a^n) \to f(a)$ for all $f \in V$, indeed for $f = 1$, $f(a^n) = 1$ for all n.

For the next definition refer to 16.0.4:

6. DEFINITION. *Let* V *be as in Remark 1. Then* V *is called a* C *space if* $\sigma(\ell, V)$ *is sequentially complete i.e.* $\{a^n\}$ *V-Cauchy implies* $f(a^n) \to f(a)$ *for all* $f \in V$. *(Remark 4)*

7. EXAMPLE. c *is not a* C *space* by Example 5. This is also a special case of Theorem 8 with $A = I$, $c_A = c = V$, $W(A) = c_o$. *However* c_o, ℓ^∞ *are* C *spaces.* These are fairly easy to check directly. In the case of ℓ^∞ one uses 1.3.7 or Example 11 or 13. For c_o it follows from 16.0.1 or by the next result and the monotonicity theorem 10.2.9 since $W(c_o) = c_o$; see also Example 11 or 13.

8. THEOREM. *With* V *as in Remark 1,* V *is a* C *space iff every matrix* A *such that* $c_A \supset V$ *has the property* $V \subset W(A)$.

Recall the definition of W in 5.6.1. Sufficiency: Let $\{a^n\}$ be V-Cauchy and $A = (a^n_k)$. Then $c_A \supset V$ (Lemma 3) and so $V \subset W(A)$. By 13.2.14, $\Lambda = 0$ on V i.e. $x \in V$ implies $\lim_A x = ax$. This is the same as $x(a^n) \to x(a)$. Necessity: Let a^n be the nth row of A. By Lemma 3 and the definition, $x(a^n) \to x(a)$ for all $x \in V$ i.e. $\Lambda = 0$ on V. By 13.2.14 it is sufficient to show that $V \subset F(A)$; this follows since $V \subset \ell^\infty = F^+(c_o)$ (10.4.7) $\subset F^+(A)$ (10.2.9) and $V \subset c_A$.

The next result is an improvement of 6.1.2.

9. THEOREM. *Let X be conull and conservative and A a matrix such that $c_A \supset W_b(X)$. Then A is conull.*

Of course A is conservative also. Suppose it is coregular. Let $D = A - \chi(A)I$. Then D is conull, hence so is $X \cap c_D$ (4.6.6) which thus includes some $\Omega(r)$ (6.4.3) which we denote by Y. In the following calculation we make several applications of the monotonicity theorem 10.2.9: $Y \cap \ell^\infty = W_b(Y)$ (13.4.15) $\subset W_b(X \cap c_D) \subset c_D \cap W_b(X) \subset c_D \cap c_A = c$ as in the proof of 3.5.4. Since Y is conull (6.3.3), this is impossible by 6.4.5 or 6.1.2.

10. THEOREM. *Let X be an FK space $\supset c_o$. Then $W_b(X)$ is a C space.*

To apply Theorem 8 assume that $c_A \supset W_b(X)$. Let $z \in W_b(X)$, $Y = z^{-1}\cdot X$. By 6.5.1, 6.5.2, Y is conull and conservative. Next we show that

$$c_{A\cdot z} \supset W_b(Y) \tag{1}$$

Let $g \in X'$ and define $f \in Y'$ by $f(y) = g(z\cdot y)$ (6.5.1). For $y \in W_b(Y)$, $g(z\cdot y) = f(y) = \sum f(\delta^k)y_k = \sum g(\delta^k)z_k y_k$ so that $z\cdot y \in W_b(X)$ with yields (1) by the definition of A. By Theorem 9, $A\cdot z$ is conull and so $z \in W_b(A)$ by 12.1.6 $(Y = c)$.

11. EXAMPLE. Taking $X = \omega$ in Theorem 10 shows that ℓ^∞ is a C space; $X = c$ shows that c_o is a C space.

12. EXAMPLE. There exists no FK space X such that $W_b(X) = c$ by Theorem 10 and Example 7.

13. EXAMPLE. *Every ideal* I *in* ℓ^∞ *such that* $I \supset c_o$ *is a* C *space.* This shows yet once more that c_o, ℓ^∞ are C spaces. Let $c_A \supset I$, $z \in I$ and $M = A \cdot z$. Then M is coercive, hence strongly conull (5.2.12). This shows that $z \in S(A)$ (12.1.7 with $Y = c$) and *a fortiori* $z \in W$. The result follows by Theorem 8.

14. Our discussion deduced the sequential completeness theorem 10 from results which are essentially equivalent to the bounded consistency theorem 5.6.8. The approach of [12] is to prove Theorem 10 directly by functional analytic methods and to deduce the bounded consistency theorem from it via Theorem 8 and 5.6.9.

15. By Example 7, $\sigma(\ell,c)$ is not sequentially complete; however $\sigma(c',c)$ is (16.0.1). Now ℓ has codimension 1 in c' (14.1.2) so ℓ *has one-dimensional sequential completion in this topology.*

16. A simple minded attempt to extend these results to sc spaces would lead to the conjecture that $\sigma(bs,V)$ is sequentially complete if $V = W_b(c_A)$. This is false even for $A = 0$. However the result of Theorem 9 can be extended; this is done in [63].

16.2. SEPARABILITY

In this section we show that the FK program can be extended

to situations in which it has failed, provided that separability assumptions are made. The contribution from functional analysis is N.J. Kalton's closed graph theorem, 16.0.3. The first result shows that this is a valid part of the FK program.

1. THEOREM. *Let* A *be a matrix then* $X = c_A$ *is a separable FK space.*

Let T_1 be the topology of ω_A restricted to X. Since ω_A is separable and metrizable (4.3.8) the same is true of (X, T_1). Let T_2 be the topology induced by $\|\cdot\|_A$ on X. Now $A[X]$ is a subspace of c hence is a separable metric space with $\|\cdot\|_\infty$. It follows that (X, T_2) is separable. Now the natural FK topology for X is $T_1 \vee T_2$ (4.3.13) and is separable (16.0.2)

2. EXAMPLE. *The FK program fails for 16.1.8.* Take $V = X = \ell^\infty$. Then $V \not\subset W(X) = c_0$.

3. THEOREM. *Let* V *be a* C *space and* X *a separable FK space. Then* $X \supset V$ *implies that* $V \subset W(X)$.

Let V be given the topology $\sigma(V, \ell)$ 16.0.4. This means that $x \to 0$ in V iff $\sum x_i y_i \to 0$ for each $y \in \ell$. The inclusion map $i: V \to X$ has closed graph since it is continuous (hence has closed graph) when each space is given the relative topology of ω and this topology is smaller than the given topologies on V, X. (For X, this is by the definition of FK space; for V it is because $x \to 0$ in ω iff $\sum x_i y_i \to 0$ for each $y \in \phi$.) By 16.0.3, i is weakly continuous. Now let $x \in V$. Then $x^{(n)} \to x$ in V as is obvious from the criterion for convergence in V just given; hence $x^{(n)} \to x$ weakly in X so $x \in W(X)$.

4. The proof of Theorem 3 does not use the full force of the assumption that $c_0 \subset V \subset \ell^\infty$. It is sufficient to replace c_0 by

ϕ. In the proof of necessity in 16.1.8 the former assumption played a role, now seen to be unnecessary. The proof of sufficiency in 16.1.8 uses $V \supset c_0$ to conclude that $a \in \ell$; it is essential there.

5. COROLLARY. *Let V be a C space containing 1. Then a separable FK space which includes V must be conull.*

By Theorem 3, $1 \in W$. (Note that c_0 is a C space which is included in the coregular separable space c; some hypothesis is required.)

The FK program may be applied to 3.5.5 which states that a coercive matrix must be conull:

6. COROLLARY. *Let X be a separable FK space $\supset \ell^\infty$. Then X is conull.*

By Corollary 5 and 16.1.7.

The same is true for 6.1.2:

7. COROLLARY. *Let Y be a conull space $\supset c_0$. If X is separable and $X \supset Y \cap \ell^\infty$ (or just $X \supset W_b(Y)$) then X is conull.*

By Corollary 5 and 16.1.10.

It was pointed out in the earlier versions that these results fail for general (non-separable) FK spaces.

8. COROLLARY. *Let V be a conservative FK space which is a C space. Then V is not separable.*

For Corollary 5 would otherwise imply that V is conull which is impossible since $V \subset \ell^\infty$ (4.6.3). (Note that c_0 is a C space so some hypothesis is required.)

9. The results of this section hold under weaker hypotheses. In [12], p. 820, separability of X is replaced by the assumption

that X has no (closed) subspace isomorphic with ℓ^∞. This yields, for example the following bigness theorem: *if Y is conull then* $X = Y \cap \ell^\infty$ *has a subspace isomorphic with* ℓ^∞ for if not, the FK space X is conull by the improvement of Corollary 7. This is impossible as in Corollary 8.

10. A.K. Snyder calls a set V *pseudo-conull* if $c_A \supset V$ implies that A is conull. For example the coregular space ℓ^∞ is pseudo-conull (Corollary 7; take $Y = \omega$). Theorem 16.1.9 shows that certain spaces W_b are pseudo-conull. Theorem 5 suggests the conjecture that any separable FK space which includes a pseudo-conull space must be conull — this is unsolved. An equivalent phrasing is the question: *does there exist a separable coregular pseudo-conull FK space?* Snyder has proved that the answer is no for BK spaces. He has also proved that if Y is conull, $Y \cap \ell^\infty$ is pseudo-conull in the weak sense that every separable space which includes it is conull. This is an extension of Corollary 7 to sc spaces.

16.3. DENSE SUBSPACES OF ℓ^∞

We show that a dense proper subspace of ℓ^∞ cannot be a bounded convergence domain. (Theorem 5.) This result is then used in the FK program, Theorem 8.

The first three results would be trivial if $D \supset c_0$.

1. REMARK. In this section D is a dense subset of ℓ^∞.

2. LEMMA. *Let A be a matrix with rows in* ℓ *such that* $\ell^\infty_A \supset D$. *Then* $\|A\| < \infty$.

If this is false there is a sequence $\{r(n)\}$ such that $\sum_k |a_{r(n),k}| \to \infty$. Let $u_n = \sum_k a_{r(n),k}$ and $b_{nk} = a_{r(n),k}/u_n$.

Then $B \in (D:c_o)$. But also $\|B\| < \infty$ so B is a continuous map of ℓ^∞ into itself, hence $B \in (\ell^\infty:c_o)$. This contradicts $B1 = 1$.

3. LEMMA $D^\gamma = D^\beta = \ell$.

Let $a \in D^\gamma$ and define A by $(Ax)_n = \sum_{k=1}^{n} a_k x_k$. By Lemma 2 $\|A\| < \infty$ which implies that $a \in \ell$.

4. LEMMA. *In Lemma 2 the assumption on the rows of A may be omitted.*

The rows of A belong to D^γ, hence are in ℓ by Lemma 3.

5. THEOREM. *Let $Y = \omega$, c_o, c or ℓ^∞, $X = Y_A$ where A is a matrix. If $X \supset D$ then $X \supset \ell^\infty$.*

Lemma 2 covers the last case. If $Y = c_o$ or c, then $\|A\| < \infty$ by Lemma 2, so $Y_A \cap \ell^\infty$ is closed in ℓ^∞ (1.3.11). Finally if $Y = \omega$, the rows of A are in ℓ by Lemma 3.

6. COROLLARY. *Let V be a dense subspace of ℓ^∞ which includes c_o. Then V is a C space.*

If A is a matrix such that $c_A \supset V$ it follows from Theorem 5 that $c_A \supset \ell^\infty$. Since ℓ^∞ is a C space (16.1.7), $V \subset \ell^\infty \subset W(A)$ (16.1.8) so V is a C space (16.1.8).

7. LEMMA. *Let V be a dense subspace of ℓ^∞ which includes ϕ. Then $\sigma(\ell,V)$ and $\sigma(\ell,\ell^\infty)$ (16.0.4) have the same compact sets.*

It is sufficient, by 16.0.7, to show that they have the same convergent sequences. Suppose that $a^n \to 0$ in the first topology. This implies that $c_A \supset V$ (16.1.3) so A is coercive by Theorem 5. This implies that $\|a^n\|_1 \to 0$ (1.7.18 ii) and so $a^n \to 0$ weakly in ℓ. The opposite implication is trivial.

8. THEOREM (G. Bennett and N.J. Kalton). *Let V be a dense*

subspace of ℓ^∞ *which includes* c_0 *and* X *a separable FK space* $\supset V$. *Then* $X \supset \ell^\infty$.

Let $x \in \ell^\infty$. Then $x^{(n)} \to x$ in $\tau(\ell^\infty, \ell)$ (16.0.6). By 16.0.8 and Lemma 7, $\{x^{(n)}\}$ is a $\tau(V, \ell)$ Cauchy sequence. Now consider the inclusion map $i: V \to X$. This is continuous, hence has closed graph, when each space has the relative topology of ω; thus it has closed graph when the larger topologies $\sigma(V, \ell)$ and the FK topology of X are used. The latter is $\tau(X, X')$ (16.0.10). By 16.0.3, 16.0.9 and Corollary 6, i is continuous with these topologies and so $\{x^{(n)}\}$ is Cauchy in X, hence convergent. It can only converge to x since coordinates are continuous. Thus, finally $x \in X$.

9. EXAMPLE. Consider this property of an FK space X:

(*): *X is the intersection of all the separable FK spaces which include it.* Every separable space has (*) as well as every X such that $X^{\beta\beta} = X$, for example, ℓ^∞, since $X = \cap\{z^\beta: z \in X^\beta\}$, and each z^β is separable by 4.3.7. However if X is an FK space $\supset c_0$ which is a dense proper subspace of ℓ^∞ (10.0.3d), X does not have (*) by Theorem 8. In [11], p. 31, it is shown that $bs + c_0$ is another example.

The space X mentioned in Example 9 shows that separable may not be omitted in Theorem 8.

10. Lemma 3, 8.6.2 and 8.6.7 suggest the following problem. *Is there an FK space* $X \supset \phi$ *with* $X \neq c_0$ *and (1)* $X \subset c_0$, *(2)* $X^\beta = \ell$? Note that (1) and $X^\beta \subset \ell$ implies (2) by 7.2.2; if X has AD, (2) implies (1) by 8.6.8. An example not satisfying (1) is given by letting E be a quasicomplement for c_0 (10.0.3d) and $X = \ell \oplus E$ (4.5.1). A much easier example takes E to be

the closure in ℓ^∞ of the periodic sequences, $X = \ell \oplus E$. See [14A], Corollary 2.

Sources: [11], [12].

CHAPTER 17
MAPS OF BANACH SPACES

17.0. FUNCTIONAL ANALYSIS

1. Let X, Y be Banach spaces, $T \in B(X,Y)$ (4.0.6), T' as in 14.0.5. Then $\|T'\| = \|T\|$; T is a linear homeomorphism (into) iff T'' is, and iff T' is onto; T' is one to one iff the range of T is dense in Y. [80], Theorems 11-3-1, 11-3-4, Corollary 11-1-8. Some of this is 7.0.5.

2. Continuing 1: T' is always weak * continuous; NT'' is the weak * closure of NT. Here NT denotes the null space of $T = \{x: Tx = 0\}$. If $y \in Y \cap T''[X'']$, then y lies in the closure of the range of T. [80], Theorem 11-1-6, Lemma 11-1-7, #11-1-107.

3. The spaces c and ℓ have no reflexive closed subspaces, except for those of finite dimension. [80], ##14-1-103, 104.

4. Let X be a normed space and $\{x_n\}$ a bounded sequence such that $f(x_n) \to 0$ for all f in a (norm) dense subset. Then $x_n \to 0$ weakly. [80] #9-3-104.

5. With X, Y, T as in 1, suppose that T is not range closed, i.e. $T[X]$ is not closed in Y. Then, for each $\varepsilon > 0$, there exists $z \in Z$ such that $\|z\| = 1$, $\|Tz\| < \varepsilon$. [80], Lemma 11-3-6.

6. Let X be a Banach space and S a subspace of X. Then S is weak * closed in X'' iff S is reflexive. [80], #11-3-118. A shorter but more advanced proof is given in [32], Lemma 1.

17.1. MATRIX MAPS OF c

We now return to the study of conservative matrices and relate summability properties of a matrix to its behavior as a map from c to c.

NOTE: *Coregular and conull matrices are now assumed conservative* as in Chapters 1-8.

1. LEMMA. *Let* $A \in \Gamma$ *be coregular,* $g \in c'$. *Then* $g = 0$ *on* $A[c]$ *iff* $g(y) = ty$ *for* $y \in c$ *with* $t \perp A$ *(3.3.1).*

Sufficiency: $g(Ax) = t(Ax) = (tA)x = 0$ (1.4.4). Necessity: We may assume that $\chi(A) = 1$. Now (3.2.1) $g(y) = \mu \lim y + ty$ so if $y = Ax$, $x \in c$, $g(y) = \mu \lim_A x + t(Ax) = \mu[\lim x + ax] + t(Ax)$ (1.3.8) $= \mu \lim x + \gamma x$ as in 4.4.9. The hypotheses is that $g(y) = 0$ whenever $x \in c$; taking $x = \delta^k$ yields $\gamma_k = 0$ so $g(y) = \mu \lim x$. Taking $x = 1$ yields $\mu = 0$ and so $g(y) = ty$ $= t(Ax)$. Taking $x = \delta^k$ yields $t \perp A$.

2. EXAMPLE. Let $(Ax)_n = x_n - x_{n-1}$, $g(y) = \lim y$. Then $g = 0$ on $A[c] \subset c_o$, but g does not have the form of Lemma 1.

3. THEOREM. *A coregular matrix is of type M iff* $A[c]$ *is dense in* c.

From Lemma 1 and the Hahn-Banach theorem (3.0.1).

4. THEOREM. *A Tauberian matrix* A *must be range closed i.e.* $A[c]$ *is a closed subspace of* c.

One might expect this since c is closed in c_A (6.1.1) and A maps these sets onto $A[c]$ and c; indeed the proof is very close to those of 3.4.4 and 6.1.1 — the difference is that the final calculation takes place in c rather than c_A.

Assume that the conclusion is false and, without loss of generality, that $\|A\| = 1$. Fix an integer $u > 1$ and set

$Z = \{x \in c_o: x_1 = x_2 = \ldots x_{u-1} = 0\}$. Then $A[Z]$ has codimension $\leq u$ in $A[c]$, hence is not closed in $A[c]$ (3.0.3). By 17.0.5 with $X = Z$, $Y = c$, for each $\varepsilon > 0$ there exists $z \in Z$ with $\|z\|_\infty = 1$, $\|Ax\|_\infty < \varepsilon$. To summarize: Given $\varepsilon > 0$ and integer $u > 1$, there exists

$$z \in c_o \text{ such that } z_k = 0 \text{ for } 1 \leq k < u, \ \|z\|_\infty = 1, \ \|Az\|_\infty < \varepsilon \quad (1)$$

Choose z^1 to satisfy (1) with $u = 2$, $\varepsilon = \frac{1}{2}$. Let $x_n^1 = z_n^1$ if $|z_n^1| > \frac{1}{2}$, otherwise $x_n^1 = 0$. Choose $n_2 = 2 + \max\{n: x_n^1 \neq 0\}$, possible since $z^1 \in c_o$.

Choose z^2 to satisfy (1) with $u = n_2$, $\varepsilon = 1/4$. Let $x_n^2 = z_n^2$ if $|z_n^2| > \frac{1}{4}$, otherwise $x_n^2 = 0$. Choose $n_3 = 2 + \max\{n: x_n^2 \neq 0\}$. Continuing in this way we get x^k satisfying: for each n, $x_n^k \neq 0$ for at most one value of k; there are infinitely many n such that $x_n^k = 0$ for all k; $\|x^k\|_\infty = 1$; $\|Ax^k\|_\infty \leq \|A(x^k - z^k)\|_\infty + \|Ax^k\|_\infty \leq \|x^k - z^k\|_\infty + 2^{-k}$ (using $\|A\| = 1$) $\leq 2^{-k+1}$. Finally let $x_i = x_i^k$ if $x_i^k \neq 0$, $x_i = 0$ if no such k exists. Then $|x_i| \leq 1$ for all k, $x_i = 1$ and 0 for infinitely many i, so x is bounded and divergent. Let $y = \sum Ax^k$. Then $y \in c$ since $\sum \|Ax^k\|_\infty < \infty$ (3.0.5) and $y = Ax$ since $y_n = \sum (Ax^k)_n = \sum_k \sum_i \{a_{ki}x_i^k: x_i^k \neq 0\} = (Ax)_n$. Thus A is not Tauberian.

17.2. MAPS OF c

Let T be a continuous linear map (called *"operator"*) from c to itself. Fix n and consider $P_n \circ T \in c'$ i.e. $(P_n \circ T)(x) = P_n[Tx] = (Tx)_n = v_n \lim x + (Ax)_n$ where $v_n = \chi(P_n \circ T)$ (3.2.1), $a_{nk} = P_n[T(\delta^k)] = (T\delta^k)_n$.

2. THEOREM. *Given T, there exist a sequence $v \in \ell^\infty$ and a matrix A with convergent columns and $\|A\| < \infty$ such that*

$Tx = v \lim x + Ax$. The matrix A need not be conservative and T is given by a matrix iff $v = 0$.

The kth column of A is $T\delta^k \in c$, also (1.0.2) $|\chi_n| + \sum |a_{nk}| = \|P_n \circ T\| \leq \|T\|$. An example is $(Tx)_n = (-1)^n [\lim x - x_n]$ in which $a_{nn} = (-1)^n$, $a_{nk} = 0$ if $n \neq k$.

3. DEFINITION. *T is called an almost matrix if $v \in c$, equivalently if A is conservative in Theorem 1.*

The equivalence is because $A \in \Gamma$ iff $1 \in c_A$ (1.3.6), hence (since $A1 \in c$) iff $v \in c$.

4. THEOREM. *Let $\chi = \chi(T) = \chi(\lim \circ T)$. Then $\chi = \lim(v+A1) - \sum a_k$. If T is an almost matrix, $\chi(T) = \lim v + \chi(A)$.*

This is clear from the definition of χ (3.2.1), and the expression for T in Theorem 2. As usual $a_k = \lim a_{nk} = \lim(T\delta^k)_n$.

5. DEFINITION. *For any vector space X, $u \in X$, g a linear functional on X, $T = u \otimes g$ (u tensor g) means $Tx = g(x)u$.*

This operator has one dimensional range (unless u or $g = 0$).

6. If T is an almost matrix, $T = v \otimes \lim + A$ where A is a matrix. Thus T is the sum of two operators on c, one being a one (or zero) dimensional operator, the other a matrix.

7. EXAMPLE. Let $g \in c'$ have the special form $g(x) = bx$, i.e. g is a row function (11.1.1). Then $u \otimes g$ is the matrix A given by $a_{nk} = u_n b_k$.

17.3. THE DUAL MAP

The use of the dual map (14.0.5) as in this chapter was suggested by J. P. Crawford and R. J. Whitley. For $X = Y = c_o$, T must be a matrix and $T' : \ell \to \ell$ is the transposed matrix. We shall determine the form of T' for $X = Y = c$. The most convenient form is as a map from ℓ to ℓ; this is available since c' is equivalent to ℓ by $h(f) = (\chi, t_1, t_2, \dots)$ where $\chi = \chi(f)$ (3.2.1), $t_n = f(\delta^n)$; see 1.0.2. Thus $S = h \circ T' \circ h^{-1}: \ell \to \ell$ may be identified with T'. Since ℓ has AK, S is a matrix and we proceed to compute it. It will be convenient to write members y of ℓ as $(y_0, y_1, y_2, \dots)$ and members x of c as $(x_1, x_2, x_3, \dots)$. Particular values of h are

$$h(\lim) = \delta^o, \quad h(P_k) = \delta^k \quad \text{for } k \geq 1 \tag{1}$$

The zero column (S_{no}) of S is $S\delta^o = h\{T'(h^{-1}\delta^o)\} = h\{T'(\lim)\}$ by (1) and this is $h(\lim \circ T)$ by definition of T'.

From the definition of h this is $(\chi, a_1, a_2, \dots)$ where these numbers are given in 17.2.4.

The kth column (S_{nk}), $k \geq 1$, is $S\delta^k = h(P_k \circ T)$ by (1) and a calculation similar to the preceding one. Thus the kth column is $(v_k, a_{k1}, a_{k2}, \dots)$ where v, A are given in 17.2.1. For example $(P_k \circ T)(\delta^n) = (T\delta^n)_k = a_{kn}$.

1. Thus S, which we identify with T' is given by the matrix

$$\begin{matrix} \chi & v_1 & v_2 & v_3 & \cdots \\ a_1 & a_{11} & a_{21} & a_{31} & \cdots \\ a_2 & a_{12} & a_{22} & a_{32} & \cdots \\ \cdots & \cdots & \cdots & \cdots & \cdots \end{matrix}$$

2. In abbreviated form, $T' = \begin{pmatrix} \chi & v \\ a & A^T \end{pmatrix}$ where the symbols are explained in 17.2.1-4. If T is a matrix, $T = A$ and $T' = A'$ $= \begin{pmatrix} \chi & 0 \\ a & A^T \end{pmatrix}$. Since ℓ has AK, the dual of a self-map (which must be a matrix) is the transposed matrix and so $T'': \ell^\infty \to \ell^\infty$ is

given by $\begin{pmatrix}\chi & a\\ v & A\end{pmatrix}$. It is very important to observe that the natural embedding of $x \in c$ into ℓ^∞ is given by $(\lim x, x_1, x_2, \ldots)$ and $y \in \ell^\infty$ lies in $\hat{c}$ iff $\lim y = y_1$, in which case $y = \hat{x}$ with $x = (y_2, y_3, y_4, \ldots)$.

3. The fact that χ is multiplicative on Γ (6.1.8) is trivial from Remark 2 since $(AB)' = B'A'$.

4. *T'' is always a matrix and is conservative iff T is an almost matrix,* in which case $\chi(T'') = \chi(A) = \chi(T) - \lim v$ by 17.2.4.

5. *If A is a conull matrix, A cannot be a linear homeomorphism of c (onto a closed subspace).* For the first row of A' is 0 and so A' cannot map ℓ onto ℓ. (See 17.0.1). In particular *no linear homeomorphism of c onto c_o can be given by a matrix.*

6. EXAMPLE. *An almost matrix which maps c one to one onto c_o.* Let $T = 1 \otimes \lim - A$ (17.2.5) where $(Ax)_n = x_{n-1}$ $(x_o = 0.)$ If $Tx = 0$ then $0 = (Tx)_1 = \lim x$; for $n > 1$, $0 = (Tx)_n = \lim x - x_{n-1}$ so $x = 0$ and T is one to one. Also the range of T is included in c_o and is all of c_o since for $y \in c_o$, $Tx = y$ where $x_n = y_1 - y_{n+1}$.

7. *No matrix can have the property of Example 6.* For such a matrix would be conull; indeed multiplicative 0, and Remark 5 applies.

8. EXAMPLE. The matrix A given by $(Ax)_n = x_{2n-1} - x_{2n}$ maps c onto c_o, while if $(Ax)_n = x_1 + x_{2n} - x_{2n+1}$, A is a conull matrix and maps c onto c.

A neater proof of 17.1.1 is now available: the condition on g is that $A'g = 0$; setting $g = (t_0,t_1,t_2,\dots) \in \ell$ this says $\chi t_0 = 0$, $a_k t_0 + (tA)_k = 0$. Thus $t_0 = 0$ and $t \perp a$ so $g(x) = t_0 \lim y + ty = ty$. The converse is similar.

17.4. w-MATRICES

The study of matrix algebras can be carried out in a very general setting which makes certain summability facts quite transparent.

1. If $T: X \to Y$ is a map of Banach spaces and T' is the dual as in the preceding section, *note that* $T''|X = T$ identifying X with its natural embedding in X''. This embedding is used to prove results in the general setting. They will hold for $X = c$ but one must be careful to observe that when specializing the actual calculations to c, the embedding of c in ℓ^∞ is not c but a subspace of c in ℓ^∞ (17.3.2).

2. REMARK. *In this section* X *is a fixed non-reflexive Banach space and* w *a fixed member of* $X''\backslash X$, Γ, ρ *are as in Definition 3.*

3. DEFINITION. $\Gamma = \Gamma(X,w) = \{T \in B(X)$: w *is an eigenvector of* $T''\}$. The corresponding eigenvalue is denoted by $\rho(T)$.

The meaning of the definition is that each $T \in B(X)$ is a continuous linear map from X to itself; $T'' = (T')'$ is as in the preceding section; $T \in \Gamma$ iff there exists a number $\rho(T)$ such that $T''(w) = \rho(T)w$.

4. EXAMPLE. Let $X = c$, $w = \chi$. This means that for $f \in c'$ $w(f) = \chi(f)$ (3.2.1). Thinking of c'' as ℓ^∞, this means that $w = \delta^0$ since $f \in c'$ corresponds to $\{\chi(f), f(\delta'), f(\delta^2), \dots\} \in \ell$

so that w picks out the first coordinate. From 17.3.2, $T''w = \{\chi, v_1, v_2, \ldots\}$ which is a multiple of w iff $v = 0$ i.e. iff T is a matrix. Thus *$\Gamma(c, \delta^o) = \Gamma$, the set of conservative matrices.* Note that $\rho(T) = \chi(T)$, but this identity will fail in more general situations; see 17.5.2. Because of the foregoing result, members of $\Gamma(X,w)$ are called *w-matrices.*

5. THEOREM. *With the notation of Remark 2, Γ is a closed subalgebra with identity of $B(X)$ and ρ is a (continuous) scalar homomorphism.*

Let $h\colon B(X) \to X''$ be given by $h(T) = T''w$. Then h is continuous since $\|h(T)\| \leq \|T''\| \cdot \|w\| = \|T\| \cdot \|w\|$ (17.0.1). Thus $\Gamma = h^{-1}[\text{span } w]$ is closed; it is a subalgebra and ρ is multiplicative since $A, B \in \Gamma$ implies $(AB)''w = A''\rho(B)w = \rho(B)\rho(A)w$. Continuity is by 4.0.12. Finally $I'' = I$ so $I \in \Gamma$.

It is trivial that the set Γ^o of conull w-matrices, those for which $\rho = 0$, is an ideal in Γ. This was noted in the classical case in 6.1.9 and 17.3.3. The next result is somewhat unexpected, even in the classical case of Example 3.

6. THEOREM. *Γ^o is a left ideal in $B(X)$.*

For $T \in B(X)$, $A \in \Gamma^o$, $(TA)''w = T''\rho(A)w = 0$.

But not a right ideal — a simple example is gotten by taking $X = c$, $T = 1 \otimes \lim$, $A = 1 \otimes P_1$ (see 17.2.5). Then $AT = T$. This also follows from general characterizations in [15].

17.5. w-ALMOST MATRICES

If T is an almost matrix, 17.3.2 shows that $T''\delta^o = (\chi-\ell)\delta^o + (\ell, v_1, v_2, \ldots)$ where $\ell = \lim v_n$. Setting $\rho = \chi-\ell$, this gives

$\rho\delta^o + \hat{v}$ where $\hat{v}$ is the natural embedding of v into $c'' = \ell^\infty$. In the next definition we identify $x \in X$ with its natural embedding in X''. When $X = c$, X'' is the second dual of c and not ℓ^∞ (17.3.2).

1. DEFINITION. *With the notation of 17.4.2,* $\Gamma_a = \Gamma_a(X,w) = \{T \in B(X): T''w \in w \oplus X\}$. *The functional* ρ *on* Γ_a *is defined by* $T''w = \rho(T)w + x$ *and* T *is called conull (coregular) if* $\rho(T)$ *is (not) zero. The members of* Γ_a *are called w-almost matrices.*

This extends the concept of coregular and conull to conservative maps which are not matrices (in the setting $X = c$, $w = \delta^o$). The development beginning in Chapter 9 extended it to matrix maps which are not conservative.

2. As pointed out before Definition 1, $\rho(T) = \chi(T) - \lim v_n$ in the classical case. Also $\rho(T) = \chi(T'') = \chi(A)$ by 17.3.4. All of these identities assume that T is an almost matrix.

3. $\Gamma \subset \Gamma_a$ *and the functional* ρ *has the same meaning on both.* In Definition 1, $x = 0$ whenever T is a w-matrix.

4. THEOREM. *With the notation of Definition 1,* Γ_a *is a closed subalgebra, with identity, of* $B(X)$ *and* ρ *is a (continuous) scalar homomorphism.*

With h as in 17.4.5, $\Gamma_a = h^{-1}[X\oplus w]$ is closed (15.0.1); it is a subalgebra and ρ is multiplicative since $(ST)''w = S''[\rho(T)w+x] = \rho(T)S''(w) + Sx = \rho(T)\rho(S)w + [\rho(T)y+Sx]$. Continuity is by 4.0.12.

5. Similar calculations show that Γ_a^o, the conull w-almost matrices, *is a left ideal in* $B(X)$ as in 17.4.6.

These subalgebras have a significant property, expressed in Theorem 7, which will follow from a preliminary lemma:

6. LEMMA. *Let* $U = \Gamma$ *or* Γ_a *(Definition 1). Let* $S \in B(X)$ *and suppose that* T *and* $ST \in U$ *with either* T *or* ST *coregular. Then* $S \in U$.

First T must be coregular by Remark 5 and 17.4.6. In the following calculation $\rho = \rho(T)$ and if $U = \Gamma$, $x = y = 0$. We have $(ST)''w = S''[\rho w+x] = \rho S''w + Sx$. Also $(ST)''w = \rho(ST)w + y$. Setting $\alpha = \rho(ST)/\rho$ we have $S''w = \alpha w + (y-Sx)/\rho$.

7. THEOREM. Γ *and* Γ_a *are left inverse closed (hence inverse closed) in* $B(X)$.

This means, for example, that if $S \in B(X)$, $T \in \Gamma$, $ST = I$ then $S \in \Gamma$. The result is immediate from Lemma 6 since I is a coregular matrix.

8. Theorem 7 should not be confused with the more difficult result that Γ is inverse closed in Φ (6.1.9).

9. EXAMPLE. For X, Y Banach spaces, a continuous linear map $T: X \to Y$ is called *weakly compact* if $T''[X''] \subset Y$. It is trivial from the definition that any weakly compact map from X to X is a conull almost matrix (for any choice of w.) It is quite easy to see that *a conservative matrix defines a weakly compact map of* c *iff it is coercive* (1.7.18).

10. ρ and χ are both defined on Γ_a. We reserved the name conull for $\rho = 0$. This is done for several reasons. Look at this nice *extension of 17.3.5 to w-almost matrices.* [If T is conull, $T''w = x$ (Definition 1) $= Tu$ for some $u \in X$ if T is range closed (17.0.2). So $T''w = T''u$, T'' is not one to one and T is not a homeomorphism (17.0.1).] But the map of 17.3.6 has $\chi = 0$. Also χ *is not a homomorphism on* Γ_a since the map of 17.3.6 has the left inverse B given by $(By)_n = y_1 - y_{n+1}$.

11. The second matrix of 17.3.8 has no right inverse in Γ_a since $\rho = 0$. It has the right inverse in $B(c)$ given by $x \to (\ell, x_1, \ell, x_2, \ell, x_3, \ldots)$ where $\ell = \lim x$. This shows that *$B(c)$ allows no non-zero scalar homomorphism.*

12. EXAMPLE. Let $V = a \otimes g$ (17.2.5). Then $\langle x, V'f\rangle = \langle Vx, f\rangle = g(x)f(a)$ so $V'(f) = f(a)g$ i.e. $V' = g \otimes \hat{a}$. Hence *$V'' = \hat{a} \otimes \hat{g}$, in particular $V''w = w(g)a$ so V is a conull w-almost matrix* and is weakly compact. *It is a w-matrix iff $w(g) = 0$.*

13. *Every $T \in \Gamma_a$ is the sum of a w-matrix and a one (or zero) dimensional operator.* This generalizes 17.2.6. Choose $h \in X'$ with $w(h) = 1$ and let $S = T - a \otimes h$ where $T''w = \rho w + a$. Then by Example 12, $S''w = \rho w + a - w(h)a = \rho w$.

14. Almost matrices were first introduced in [78] in the way just described. They later appeared independently as the solution of a problem concerning commuting operators. For example, every $T \in B(c)$ which commutes with $(C,1)$ must be an almost matrix. See [54] p. 213.

In [20] and articles cited there, the problem is studied as to whether, for various choices of w, the algebras $\Gamma(X,w)$ are isomorphic, and the same question for Γ_a. The answer is yes for $X = c$, no for $X = \ell$.

16. In [29] the two algebras are introduced without use of dual maps and in the more general setting of locally convex spaces which are not weakly complete. For example Γ may be defined by means of a weakly Cauchy and divergent net w in X, namely $\{T : (T-\rho I)(w) \to 0 \text{ weakly}\}$.

17.6. TAUBERIAN MAPS

1. DEFINITION. *Let* X, Y *be Banach spaces and* $T: X \to Y$ *continuous and linear;* T *is called Tauberian if* $T''w \notin Y$ *whenever* $w \notin X$.

This is behavior exactly opposite to that of a weakly compact map, 17.5.9.

2. With the notation of 17.5.1, T is Tauberian iff for every w, T is not conull.

3. EXAMPLE. *A conservative almost matrix is Tauberian in the sense of Definition 1 with* $X = Y = c$ *iff its matrix part* A *is Tauberian.* Necessity: In particular $\hat{c}$ does not contain $T''\delta^{o}$ $= (\chi, v_1, v_2, \ldots)$. This means that $\lim v \neq \chi$ i.e. $\rho(T) \neq 0$ (17.5.2). Now let $x \in c_A \cap \ell^{\infty}$, $u = (\lim_A x - ax)/\rho$, $y = (u, x_1, x_2, \ldots)$. Then by 17.3.2 $(T''y)_{n+1} = v_n u + (Ax)_n \to (\chi-\rho)u + \lim_A x = \chi u + ax$ $= (T''y)_1$. Thus $T''y \in \hat{c}$ and so by hypothesis $y \in \hat{c}$ hence $x \in c$. (Also $u = \lim x$. This was the motivation for our definition of u, using 1.3.8.) Sufficiency: A is coregular, (3.5.2), hence T is (17.5.2). Suppose $T''w \in \hat{c}$. Let $x = (w_2, w_3, w_4, \ldots)$. Then $v_n w_1 + (Ax)_n = (T''w)_{n+1} \to (T''w)_1 = \chi w_1 + ax$. Since $v_n \to \chi - \rho$ it follows that $Ax \in c$ and $\lim_A x = \rho w_1 + ax$. By hypothesis, $x \in c$ and so $\lim_A x = \rho \lim x + ax$ (1.3.8). Hence $\lim w = \lim x = w_1$ so $w \in \hat{c}$ (17.3.2).

By Example 3, Tauberian maps generalize Tauberian matrices. Their study has led to some interesting problems in general Banach space theory. We shall do only enough of the general theory to cover the classical characterization, Theorem 10.

4. THEOREM. *The null space of a Tauberian map must be reflexive.*

If $T''F = 0$, $F \in X''$, the hypothesis implies that $F \in X$ i.e. $NT'' = NT$. ($NT = \{x : Tx = 0\}$ is the null space of T.) It follows that NT is weak * closed in X'' (17.0.2) and so it is reflexive (17.0.6).

The converse is obviously false since there are non-Tauberian triangles such as $(C,1)$. These are one to one maps! There is a converse result for range closed operators:

5. THEOREM. *Let T be range closed. Then T is Tauberian iff NT is reflexive.*

Half is by Theorem 4. Conversely, let $w \in X''$, $T''w \in Y$. By 17.0.2, $T''w = Tx$ for some $x \in X$ so $w-x \in NT''$ which is the weak * closure of NT (17.0.2). But NT is weak * closed in X'' so $w-x \in NT$. In particular $w \in X$.

6. LEMMA. *The composition of two Tauberian maps is Tauberian.*

If $(TS)''w \in Z$ then $T''(S''w) \in Z$ so $S''w \in Y$ and $w \in X$.

7. COROLLARY. *The restriction of a Tauberian map to a closed subspace is Tauberian.*

Let $X \subset Y$ and $T: Y \to Z$ Tauberian. Then $T|X$ is the composition of T with the inclusion map. The latter is Tauberian by Theorem 5 and the result follows by Lemma 6.

8. DEFINITION. *A semi-Fredholm map is a range closed map whose null space is finite dimensional.*

Thus a semi-Fredholm map must be Tauberian, (Theorem 5). The converse is false; for every map of a reflexive space is Tauberian (and weakly compact).

9. EXAMPLE. *Let* $T\colon \ell \to Y$, Y *any Banach space. Then* T *is Tauberian iff it is semi-Fredholm.* Sufficiency is obvious by Theorem 5. Conversely, the null space of T is finite dimensional by Theorem 4 and 17.0.3. Thus ℓ has a closed subspace X on which T is Tauberian (Corollary 7) and one to one and such that $T[X] = T[\ell]$. Denoting $T|X$ by T it is sufficient to show that $T\colon X \to Y$ is range closed. If not there is a sequence $\{x_n\}$ in X with $\|x_n\| = 1$, $Tx_n \to 0$ (17.0.5). We shall prove that $x_n \to 0$ weakly in X, hence in ℓ, and this contradicts 1.7.20. First T'' is one to one since $NT'' = NT$ as in the proof of Theorem 4. Thus $T'[Y']$ is norm dense in X' (17.0.1). Also $f \in T'[Y']$ implies $\langle x_n, f\rangle = \langle x_n, T'g\rangle = \langle Tx_n, g\rangle \to 0$ and so $x_n \to 0$ weakly by 17.0.4.

The classical theorem given next is due in part and in full with various proofs to I.D. Berg, J.P. Crawford, R.J. Whitley and to S. Mazur and W. Orlicz. The proof given here is that of [32]. Other references are given there. An easy proof for row-finite matrices is given in [82], Remark 4.3. The more general result that *if* X *has no reflexive subspace, Tauberian = semi-Fredholm for maps:* $X \to Y$ is given in [40], 4.3.

10. THEOREM. *A conservative matrix is Tauberian iff it is a semi-Fredholm map from* c *to* c.

This follows from Example 3, Theorem 5 and 17.1.4 along with the fact that c has no infinite dimensional reflexive subspace (17.0.3).

11. COROLLARY. *Let* A *be a conservative triangle. Then* A *is Tauberian iff it is range closed, Mercerian iff it is onto.*

These refer to the map $A\colon c \to c$. The conditions hold since A is one to one.

12. A Tauberian matrix A maps c onto a closed subspace of c (17.1.4). It follows from Theorem 10 that A preserves *all* closed linear subspaces. (See [82], Theorem 2.3.)

Reference: [80], pp. 175-177, 14-1-304, 305.

17.7. MULTIPLICATIVE ABSTRACTED

In the preceding sections we considered eigenvalues of T''. An obvious modification is to do the same for T'. Actually, more than idle curiosity suggests this; it is motivated by applying to multiplicative matrices the process of generalization which led from matrices to w-matrices. We shall show the abstract formulation first.

1. DEFINITION. *Let* X *be a Banach space and fix* $h \in X'$, $h \neq 0$. *Let* $M = M(X,h) = \{T \in B(X): h$ *is an eigenvector of* $T'\}$. *For* $T \in M$ *write* $T'h = m(T)h$.

2. LEMMA. *For each* $T \in M$, $h(Tx) = mh(x)$ *for all* x; $m = m(T)$.

For $\langle Tx,h\rangle = \langle x,T'h\rangle = \langle x,mh\rangle = m\langle x,h\rangle$.

3. LEMMA. $T \in M$ *iff* $T[h^\perp] \subset h^\perp$.

Necessity is by Lemma 2. Sufficiency: If $h(x) = 0$, $(T'h)(x) = \langle x,T'h\rangle = \langle Tx,h\rangle = 0$ and so $T'h$ is a multiple of h (15.0.5).

That M is a closed subalgebra, with identity, of $B(X)$ and that m is a scalar homomorphism are proved like 17.5.4.

To discuss multiplicative *matrices* in the general setting we shall *assume that* $w(h) = 1$ where w is as in 17.4.2. Let $M\Gamma$ and $M\Gamma_a$ refer to $M \cap \Gamma$ and $M \cap \Gamma_a$ respectively.

4. THEOREM. *Let* $T \in M\Gamma_a$. *Then* $m(T) = \rho(T) + h(x)$ *where*

$T''w = \rho w + x$. *Thus* $M\Gamma_a$ *has two distinct scalar homomorphisms; they agree on* $M\Gamma$ *on which they are equal to* χ.

Since $w(h) = 1$ we have $m = m\langle h,w\rangle = \langle mh,w\rangle = \langle T'h,w\rangle = \langle h,T''w\rangle = \rho\langle h,w\rangle + \langle h,x\rangle = \rho + h(x)$.

In the classical case this says that a multiplicative-m matrix has $\chi = m$. ($\rho = \chi$ on Γ).

5. EXAMPLE. Let $X = c$, $h = \lim$. Then $T \in M$ iff T is multiplicative in the summability sense i.e. $\lim Tx = m \lim x$ for $x \in c$. This is by Lemma 2. We may also take $h = \delta^0 \in \ell$ so that $T \in M$ iff $T'h = \chi h$ i.e. iff $a = 0$, from 17.3.1.

6. This formulation explains an old paradox in summability theory. The following list of facts is at first glance somewhat puzzling:

(a) The set of conull matrices in a left ideal in $B(c)$

(a') The set of multiplicative-0 operators is a right ideal in $B(c)$

(b) The set of matrices is left inverse closed in $B(c)$

(b') The set M is right inverse closed in $B(c)$.

We shall prove these two pairs of facts in the general setting and it will be seen that the puzzling shift from left to right in each pair is simply explained by the fact that matrices are defined in terms of T'' (§17.4) while multiplicative operators are defined in terms of T'; this with the facts that $(ST)'' = S''T''$, $(ST)' = T'S'$ explains the shift from one side to the other.

Fact (a) is 17.4.6; Fact (b) is 17.5.7. To prove (a)' let $A \in M$ with $m(A) = 0$, $B \in B(X)$. Then $(AB)'h = B'(A'h) = B'0 = 0$. To prove (b)' let $A \in M$, $B \in B(X)$, $AB = I$. Then $h = I'h = (AB)'h = B'(A'h) = B'(mh) = mB'(h)$, hence $B'(h) = (\frac{1}{m})h$ since

$m = m(A) \neq 0$ by (a)'.

7. Every $A \in \Gamma$ is the sum of a member of $M\Gamma$ and a one (or zero) dimensional member of Γ. This is analogous to 17.5.13 and the method is an abstraction of the trick used in 3.5.1, of which the present result is a generalization. Let $g = A'h - \chi h$. Choose x with $h(x) = 1$. (In the classical case $h = \lim$, $x = 1$ will do.) Let $B = A - x \otimes g$. Then $B \in \Gamma$ since $B''w = A''w - w(g)x - A''w$ since $w(g) = \langle A'h-\chi h,w\rangle = \langle h,A''w\rangle - \chi\langle h,w\rangle$; $A''w = \chi w$ and $\langle h,w\rangle = 1$. Also $B \in M$ since $B'h = A'h - h(x)g = A'h - g = \chi h$. Note also that $\chi(B) = \chi(A)$.

In [15] it is proved that Γ_a is a maximal proper subalgebra of $B(c)$, that χ is the only non-zero scalar homomorphism on $M\Gamma$ and ρ is the only one on Γ_a; $X = c$ in both cases. It follows that there is no non-zero scalar homomorphism on $B(X)$ if $X = c$ (this was already observed in 17.5.11); this is true also for $X = c_o$, ℓ^p and Hilbert space. See [19], [36]. However if $X'' = X \oplus w$ ([80], Remark 3-2-7) it is clear that $\Gamma_a = B(X)$ and ρ is a scalar homomorphism.

CHAPTER 18
ALGEBRA

18.0. FUNCTIONAL ANALYSIS

1. Let S be a subspace of c_0 which is linearly homeomorphic with c_0. Then S is complemented in c_0 [80], Corollary 9-6-5, #9-6-110. (Note. It is proved in [89] that no other Banach space has this property.)

2. (a) The set G of invertible members of a Banach algebra with identity is open. [79], 14.2 Fact iii. (b) Every member z of ∂G (18.2.1) is a topological divisor of 0 i.e. for each $\varepsilon > 0$ there exist x, y with $\|x\| = \|y\| = 1$, $\|xz\| < \varepsilon$, $\|zy\| < \varepsilon$. [79], #14.2.39. (c) $\overline{G}$ is closed under multiplication. This is immediate from continuity of multiplication (1.0.6).

3. Let z belong to a Banach algebra with identity. Then $\sigma(z)$, the spectrum of z, = {t: z-t1 is not invertible}. This is a set of scalars. If $t \in \sigma(z)$, then $|t| \leq \|z\|$. [79], Theorem 14.2.3.

4. A weakly compact (17.5.9) member of B(c) must be compact [80], ##11-4-108, 11-4-203. The spectrum of a compact map of a Banach space is countable [73] p. 281 Th. 5.5G.

5. Let f be a non-zero scalar homomorphism on a Banach algebra with identity. Then $f(z) \in \sigma(z)$ for each z [79], Lemma 14.1.5.

18.1. TOPOLOGICAL DIVISORS OF 0

In this chapter we study the Banach algebras Γ of conservative matrices, and Δ of conservative triangular matrices (1.5.4, 1.5.5). An extension to sc matrices is counter-indicated by 9.5.5. We saw a relation between Mercerian (or Tauberian) and invertible in Δ (1.7.14) (or in Γ, 1.8.4).

I.D. Berg suggested the approach of this section. The original motivation was the observation of 18.2.4 which suggested the result of 18.2.5.

1. There is a conceptual difference between a matrix A and the operator $x \to Ax$. For $x \in c$, $A \in \Gamma \subset B(c)$ they are the same. This is expressed by $AB = A \circ B$, the first being matrix multiplication. See 1.4.4, 1.4.6.

2. DEFINITION. *Let* z *be a member of a Banach algebra. Then* z *is called an* ℓtz *(left topological divisor of 0) if for each* $\varepsilon > 0$ *there exists* y *with* $\|y\| = 1$, $\|zy\| \le \varepsilon$.

The next result is an adaptation of a theorem of B. Yood to Γ (17.4.2):

3. THEOREM. *With the notation of 17.4.2, let* $T \in \Gamma$. *Then* T *is an* ℓtz *iff* T *is not a linear homeomorphism (into).*

Necessity: Say $T: X \to S$ is a bijection where S is a closed subspace of X. Let $U = T^{-1}: S \to X$. Then for all $V \in B(X)$, $\|V\| = \|UTV\| \le \|U\| \cdot \|TV\|$ which implies that $\|TV\|$ is bounded away from 0 i.e. T is not an ℓtz.

Sufficiency: Let $\varepsilon > 0$. Choose $a \in X$ with $\|a\| = 1$, $\|Ta\| < \varepsilon$. (By 17.0.6 if T is one to one, otherwise we can make $Ta = 0$.) Let $g \in X'$, $\|g\| = 1$, $w(g) = 0$ and set $V = a \otimes g$ (17.2.5). Then for $y \in X$, $\|Vy\| = |g(y)|$ so $\|V\| = 1$, $\|TVy\| =$

$|g(y)| \cdot \|Ta\| \leq \varepsilon \|y\|$ so $\|TV\| \leq \varepsilon$. Finally $V \in \Gamma$ since (17.5.12) $V''w = w(g)a = 0$.

4. Since this proof works for any $T \in B(X)$, the same result holds for any subalgebra, such as Γ_a (17.5.1), which includes Γ.

5. THEOREM (I.D. Berg). *Let A be a conservative matrix which is one to one on c. Then A is Tauberian iff it is not an ℓtz in Γ*

From Theorem 3 and 17.6.10.

6. THEOREM. *Let A be a conservative triangle. Then A is Tauberian iff A is not an ℓtz in Δ (1.5.5).*

Necessity is as in Theorem 5. Conversely if A is not Tauberian it is not ranged closed by 17.6.10; the proof of Theorem 3 provides a matrix $V = a \otimes g$ with $\|V\| = 1$, $\|AV\| \leq \varepsilon$. By 17.2.7, $v_{nk} = a_n b_k$ where $g(x) = bx$. (g has this form since $0 = w(g) = \chi(g)$.) Let $b_{nk} = v_{nk}$ for $k \leq n$, 0 for $k > n$. Clearly $B \in \Delta$, $\|B\| = 1$ and $\|AB\| = \|AV\| \leq \varepsilon$.

7. A proof of Copping's theorem (6.1.5) follows: if $A \in \Gamma$ has a left inverse in Φ it is not an ℓtz, so is Tauberian. Also the result of 1.8.4 can be given in full generality: *if $T \in B(X)$ has a left inverse* it is not an ℓtz, hence is a linear homeomorphism by Remark 4 applied to $B(X)$. By 17.6.5 *it is Tauberian.*

8. LEMMA. *Let $T \in B(c)$ (or Γ or Γ_a). Then T is a linear homeomorphism iff T has a left inverse in $B(c)$ (or Γ or Γ_a).*

Sufficiency is obvious. Necessity: Let $U = T^{-1}: T[c] \to c$. By 18.0.1, there is a continuous projection P of c onto $T[c]$. Let $S = U \circ P$. Then for $x \in c$, $STx = U[PTx] = UTx = x$. The results for Γ, Γ_a follow from 17.5.7.

9. For $T \in \Gamma$ in Lemma 8, the left inverse map S is also a left inverse matrix i.e. $ST = I$, by 1.4.4. This is the same as Remark 1.

10. THEOREM. *With the assumptions of Theorem 5, A is Tauberian iff it has a left inverse in* Γ.

Sufficiency: Remark 7. Necessity: Theorems 3, 5 and Lemma 8.

For general $A \in \Gamma$ we have that A is Tauberian iff there exists $B \in \Gamma$ with $BA = I + D$, D a *finite* member of Γ i.e. D has only a finite set of non-zero columns. This is accomplished by adding rows to the top of A to make it one to one on c; possible by 17.6.10.

11. J. Copping, [24] Lemma 1, has an interesting extension: *Suppose that A is coregular, $B \in \Phi$ and $BA \in \Gamma$. Then there exists $D \in \Gamma$ with $DA = BA$.* We have been concerned up till now with the special case $BA = I$, in which case A is automatically coregular since it is Tauberian (Remark 6, 3.5.2). It is trivial that *if A is of type M then $B \in \Gamma$* since $(D-B)A = 0$. This is a *left inverse closure* result.

Sources: [14], [23], [77].

18.2. THE MAXIMAL GROUP

1. DEFINITION. *Let Z be a Banach algebra with identity. The maximal group $G = G(Z)$ is the set of invertible members of Z; ∂G is the boundary of G i.e. $\overline{G} \setminus G$.*

This agrees with the usual definition of boundary: $\overline{G}\backslash G^i$, since G is open (18.0.2a). Note that $a \in G$ iff there exists $b \in Z$ with $ab = ba = 1$.

2. THEOREM. *A matrix* $A \in G(\Delta)$ *(1.5.5) iff it is a Mercerian triangle.*

Necessity: A is triangular and has an inverse matrix, so it is a triangle. It is Mercerian by 1.7.14. Sufficiency is by 1.7.14.

3. THEOREM. *If* $A \in G(\Gamma)$ *it must be Tauberian.*

By 1.8.3 (or 18.1.7). It was pointed out in 1.8.6 that A need not be Mercerian.

4. So far the discussion has been easy and lacking in surprises. The interest lies in looking at ∂G. The first hint that it contains objects of interest came with the discovery of Mercer's theorem 2.4.1. In view of Theorem 2 it says that $\alpha I + (1-\alpha)C \in G$ for $\alpha > 0$. Letting $\alpha \to 0$ we see that $C \in \partial G$. For $\alpha < 0$ this matrix is Tauberian (2.4.2). Similarly 1.8.6 shows that if $(Ax)_n = x_n - \alpha x_{n-1}$, A is Mercerian for $|\alpha| < 1$, Tauberian for $|\alpha| > 1$, (replace α by $1/\alpha$ in the first matrix given in 1.8.6,) and not Tauberian for $|\alpha| = 1$. (It sums $\{\alpha^n\}$ which is divergent unless $\alpha = 1$ in which case A is conull and 3.5.2 applies.) *This suggests the conjecture that the non-Tauberian matrices are precisely those in* ∂G. The conjecture turns out to be false but is close enough to the truth to make it interesting. To avoid trivialities we add the assumption that A is one to one on c; for example the identity matrix with one diagonal member replaced by 0 is in ∂G but is Mercerian.

Half of the conjecture is true:

5. THEOREM. *Let* $A \in \Gamma$ *be one to one on* c. *If* $A \in \partial G(\Gamma)$ *it is non-Tauberian.*

By 18.0.2(b) and 18.1.5.

The extreme non-Tauberian matrices give evidence for the converse:

6. EXAMPLE. *Every coercive matrix* $A \in \partial G(\Gamma)$. A is weakly compact (17.5.9), hence compact (18.0.4). Hence there exists a sequence $u_n \to 0$ such that $A - u_n I \in G$ for each n (18.0.4). Since $A \notin G$ (Theorem 3), the result follows.

7. EXAMPLE. *Let* A *be a type* M *triangle (3.3.3). Then* A *is Tauberian iff* $A \in G(\Gamma)$. (This is the same as $A \in G(\Delta)$ in this case.) Sufficiency: Indeed A is Mercerian. Necessity: A is coregular (3.5.2), hence perfect (3.3.4). Since c is closed in c_A (6.1.1) it follows that A is Mercerian.

8. Example 7 and Remark 4 suggest the easy looking question: *find a type* M *triangle not in* $\overline{G(\Delta)}$. A good deal of effort has gone into this but the question is still open.

9. EXAMPLE. *The conjecture in Remark 4 is false.* J. Copping [23], p. 193, shows that if $(Ax)_n = x_n - 99x_{n-1} - 100x_{n-2}$ then the (essentially regular) triangle A is an interior point of the set of non-Tauberian matrices in Δ. So $A \notin \partial G(\Delta)$. In connection with Remark 8, A is not of Type M since $\{10^{-2n}\} \perp A$. Copping's example is amplified in [14].

10. This form of the question in Remark 4 is still open: *must all non-Tauberian matrices lie in* $\partial G(\Gamma)$? (Again, assume one to one on c.)

11. EXAMPLE. (J. DeFranza and D.J. Fleming). *A regular non-perfect triangle in* $\partial G(\Delta)$. Let $(Ax)_n = x_{n-1} + b_n x_n$ with $b_n \to 1$. (Actually $\frac{1}{2}A$ is regular. Sue me!) If we define A_m by replacing each b_n, $n > m$, by $1 + 1/m$ we shall have $A_m \in G(\Delta)$ as in Remark 4. Also $A_m \to A$. To make A not perfect we simply choose $u_n = n^{-2}$ in 3.3.12, taking account of 3.3.4. Earlier examples [40A], [62] had A conull (and not of type M).

12. B.E. Rhoades [53] has located many of the classical matrices (Chapter 2) in $\partial G(\Delta)$; see also [21], p. 65. Naturally some Nörlund matrices are Tauberian (2.6.8, also [14], Lemma 4) so they cannot be there, by Theorem 5. Some of the questions raised in [53] are answered in [40A], [62] and, of course, Example 11.

13. Membership in $G(\Delta)$ is not invariant, for example if $(Ax)_n = x_{n-1}$, A is equipotent with I but $A \notin G$. Even more simply, replace a diagonal element in I by 0; this yields a matrix in ∂G which shows that this set is also not invariant. However *membership in these sets is invariant for triangles:* this is trivial for G by Theorem 2. Next let $A \in \partial G$, $c_B = c_A$. Then $B = MA$ where $M = BA^{-1}$ is Mercerian (1.7.5) hence in G by Theorem 2. Since $\overline{G}$ is closed under multiplication (18.0.2c) it follows that $B \in \overline{G}$; finally $B \notin G$ by Theorem 2. (This argument is due to B. E. Rhoades).

Source: [77].

18.3. THE SPECTRUM.

1. Any operator T on a Banach space X has a spectrum (18.0.3) as a member of $B(X)$. If T belongs to some closed subalgebra Z with identity its spectrum as a member of Z might

a priori be larger. This cannot happen in the cases of interest here: *If* $T \in \Gamma$ *(17.4.3), its spectrum is the same computed in* Γ *or in* $B(X)$. This is by 17.5.7 which also yields a similar statement for Γ_a. The same is true for Δ (1.5.5) vis a vis Γ (1.5.1) and $B(c)$.

2. Any question about membership in a spectrum can be resolved if it is known how to tell when 0 is in a spectrum. This is because $t \in \sigma(z)$ iff $0 \in \sigma(z-t1)$. Thus all comes to deciding which objects are invertible. For example, if $A \in \Delta$ (1.5.1) A is invertible iff it is a Mercerian triangle (1.7.14).

3. LEMMA. *Fix* n *and define* $f(n) = a_{nn}$ *for* $A \in \Delta$. *Then* f *is a non-zero scalar homomorphism.*

4. COROLLARY. *For* $A \in \Delta$, *each* $a_{nn} \in \sigma(A)$.

By Lemma 3 and 18.0.5.

5. EXAMPLE. It is proved in [85A], Theorem 2, that if $A \in \Delta$ is coercive, $\sigma(A) = \{0\} \cup \{a_{nn}\}$.

6. EXAMPLE. Let C be the (C,1) matrix (1.3.10). Fix $t \neq 1$ and let $A = C-tI$. Then A is a multiple of $\alpha I + (1-\alpha)C$ with $\alpha = t/(t-1)$. Theorems 2.4.1, 2.4.2 show that the latter matrix is a Mercerian triangle iff $\alpha > 0$. The proof of 2.4.2 actually works for complex α and given the same result for $R\alpha > 0$. The details may be seen in [34], p. 106, Theorem 52. Thus A is a Mercerian triangle iff $R\alpha > 0$ i.e. (in the complex plane) iff α is closer to 1 than it is to -1 which holds iff $\left|\frac{\alpha+1}{\alpha-1}\right| > 1$. Now $t - \frac{1}{2} = \frac{1}{2}\frac{\alpha+1}{\alpha-1}$ so, finally, A is invertible iff $|t - \frac{1}{2}| > \frac{1}{2}$. *We have proved that* $\sigma(C) = \{t: |t - \frac{1}{2}| \leq \frac{1}{2}\}$. The value $t = 1$ was excluded, but is in the spectrum by Corollary 4,

or because in that case A is conull.

7. EXAMPLE. N.K. Sharma [62A] shows that for $A \in \Delta$ the only possible isolated points of $\sigma(A)$ are its diagonal elements. In particular 0 cannot be an isolated point of the spectrum of a triangle.

These examples show that it is fairly difficult to compute the spectrum of a particular matrix. The spectra of Hausdorff and weighted mean matrices are considered in [21], [62] and [62A]. The spectrum of the Hilbert matrix in $B(\ell^2)$ is considered in [22B], Note, p. 307. In [27A], Theorem 3, the spectrum of a certain Mercerian Nörlund matrix with $a_{nn} = 1$ is shown to contain negative numbers.

It is curious that the behavior of A outside c can be consulted to learn about its spectrum e.g. if $A \in \Delta$ maps some divergent sequence into c, A is not Mercerian hence $0 \in \sigma(A)$.

8. We now turn to a generalization of Γ and Γ_a (17.4.3, 17.5.1) in which X is a Banach space and w is a linear functional on X', not necessarily in X''. (Suggested by D. Franekic.) See also 17.5.16. It is unknown whether these are closed subalgebras of $B(X)$ in this case; however they are inverse closed (like 17.5.7) and this is sufficient to imply that every scalar homomorphism is continuous. The continuity of ρ will be proved directly, however (Corollary 13).

9. DEFINITION. *Let $T \in B(X)$; then λ is called an approximate eigenvalue of T if for every $\varepsilon > 0$ there exists x with $\|x\| = 1$, $\|Tx-\lambda x\| < \varepsilon$.*

For example any eigenvalue is an approximate eigenvalue and each approximate eigenvalue lies in the spectrum since $T-\lambda I$ is not range closed (17.0.5).

10. EXAMPLE. Let T be a linear homeomorphism of X onto a proper subspace. Then T is not invertible so $0 \in \sigma(T)$, but 0 is not an approximate eigenvalue.

11. THEOREM. *Let* $T \in \Gamma_a$ *(Remark 8). Then* $\rho(T)$ *is an approximate eigenvalue of* T.

Let $U = T - \rho(T)I$. Then $U''w = x \in X$. We shall show that $x \in \overline{RU}$. Let $g \in X'$, $g = 0$ on RU. Then $U'g = 0$ and so $g(x) = U''w(g) = w(U'g) = 0$. The Hahn-Banach theorem implies the assertion. If U is not range closed the conclusion of the Theorem is true by 17.0.5. If U is range closed, $x \in RU$, say $x = Uz$. Then $U''w = U''z$ and so U'' is not one to one. Hence U' is not onto (17.0.1) and U is not a linear homeomorphism (17.0.1) so the conclusion follows again.

12. EXAMPLE. Let A be a conull triangle. Then $\rho(A) = 0$ is not an eigenvalue of A since A is one to one.

13. COROLLARY. ρ *is continuous.*

By Theorem 11, $\rho(T) \in \sigma(T)$, hence $|\rho(T)| \leq \|T\|$ (18.0.3).

CHAPTER 19
MISCELLANY

19.0. FUNCTIONAL ANALYSIS

1. Every separable quotient of ℓ^∞ is reflexive [80] #15-3-1.

2. Let $\{X_n\}$ be a decreasing sequence of FK spaces and $X = \cap X_n$. Suppose that X is a dense proper subset of each X_n; then X is not a BK space. (X is an FK space by 4.2.15.) [79] #11.3.27.

19.1. WHAT CAN Y_A BE?

1. We have seen (16.2.1) that c_A is separable and this leads to inequivalence results such as that ℓ^∞ cannot be c_A. Also *c_A' with its strong topology is separable*, [8] Theorem 5, and so, for example, ℓ cannot be c_A or even a closed subspace of c_A. The first was proved in a different way in 12.4.9. Some other inequivalence results are given in 13.3.2. Also if X is a conservative FK space such that $X \cap \ell^\infty$ is separable but not equal to c, then X cannot be c_A; this follows from 6.5.6. In general an inequivalence theorem results from every failure of the FK program e.g. a coregular FK space $\supset \ell^\infty$ cannot be c_A. Finally we mention that if the codimension of $\overline{X^\beta}$ in X' is more than one, then X cannot be c_A; this follows from 14.2.3 and 14.3.1.

2. EXAMPLE. Any z^β e.g. $cs = 1^\beta$, can be c_A. Just let every row of A be z. Also c and c_0 can be c_A. For c_0, take $(Ax)_{2n} = 0$, $(Ax)_{2n-1} = x_n$. Finally every FK space of the

form $c \oplus [x^1, x^2, \ldots, x^n]$ where the x^i are linearly independent (mod ℓ^∞) can be c_A. This is proved in [85], Theorem 3, with a correction on p. 386.

3. A different sort of inequivalence theorem is that *c_A cannot be a closed subspace of ℓ*, unless it is finite dimensional (as in 1.2.6). For c_A' would then be a quotient of ℓ^∞ (17.0.1) and separable, as mentioned in Remark 1. This makes it reflexive by 18.0.1 so c_A is a reflexive subspace of ℓ, hence finite dimensional (17.0.3).

4. The general problem of characterizing which FK spaces have the form c_A seems extremely difficult.

5. EXAMPLE (G. Bennett and G. Meyers). *A matrix A such that $\ell_A^\infty = \ell^p$.* We shall show this for $p = 1$. The more general result is similar. The matrix A has $\pm\delta^1$ as its first two rows, $\pm\delta^1 \pm \delta^2$ as the next four, $\pm\delta^1 \pm \delta^2 \pm \delta^3$ as the next eight and so on. So A is row-finite and all its terms are ± 1 or 0. For $x \in \ell$, $|(Ax)_n| = |\sum (\pm x_k)| \leq \|x\|_1$ so $\|x\|_A \leq \|x\|_1$. Conversely given $x \in \ell_A^\infty$ fix any m; then $\sum_{k=1}^{m} |x_k| = \sum (\pm x_k) = (Ax)_n \leq \|x\|_A$ so $\|x\|_1 \leq \|x\|_A$.

6. With A as in Example 5, c_A is a closed subspace of ℓ (4.3.14), hence finite dimensional as pointed out in Remark 3. Actually it is clear that $c_A = \{0\}$. The map $A\colon \ell \to \ell^\infty$ cannot come close to being onto of course; it does not even cover c as just mentioned, or even ϕ. It is one to one and $y = Ax$ implies that $x_n = y_{2^n - 1} - y_{2^{n-1} - 1} = (By)_n$ say. Thus $x = By = B(Ax)$ for $x \in \ell$. Since $\ell \supset \phi$ this shows that $BA = I$. Also $A(By) = y$ for $y \in A[\ell]$, but AB is nothing like I. (Yet it is idempotent since $ABAB = AIB = AB$.)

7. We now turn to ω_A. This space has AK (4.3.8) so cannot be c, bv, ℓ^∞, bs etc. It is a β dual, namely R^β, where R is the set of rows of A, so it cannot be c_o; c_o is not a β dual since $c_o^{\beta\beta} \neq c_o$ (7.2.2).

8. THEOREM. *ω_A is a BK space iff it is E^β where E is a finite set of sequences not all in ϕ. Here E may be taken to be the first m rows of A.*

Necessity: Let E_n be the first n rows of A. Then $\{E_n^\beta\}$ is a decreasing sequence of AK spaces (4.3.7, 4.2.15) and ω_A is their intersection. It is dense in each E_n^β since it includes ϕ. It follows from 19.0.2 that $\omega_A = E_m^\beta$ for all sufficiently large m. Sufficiency: By 4.3.7.

9. EXAMPLE. Let $a_{nk} = 1$ for all n, k. Let $b_{nk} = 1$ for $n = 1$, 0 otherwise. Then $\omega_A = \omega_B = cs$. Theorem 8 shows that this situation is typical; if ω_A is a BK space there must be m such that $a_n^\beta \supset E_m^\beta$ for all n, where a_n is the nth row of A, E_m is the first m rows. (Also $a_n \in E_m^{\beta\beta}$).

10. It seems fairly obvious from Theorem 8 that $\omega_A \neq \ell$. Otherwise $\ell = E^\beta$ with E a finite set, so $\ell^\infty = E^{\beta\beta} = (\text{Span } E)\cdot bv + \phi$ which is impossible. (The last mentioned identity is not too hard to prove.)

19.2. TOEPLITZ BASIS

1. A *biorthogonal system* is a sequence of points $\{b^n\}$ and of continuous linear functionals $\{f_n\}$ such that $f_n(b^k) = \delta_n^k$ i.e. 1 if $n = k$, 0 otherwise. The most familiar such system is $\{\delta^n\}$, $\{P_n\}$. It is a *basis* for X if every $x \in X$ is $\sum f_k(x)b^k$; thus an FK space has AK iff $(\{\delta^n\},\{P_n\})$ is a basis.

2. Henceforth we shall refer to $\{b^n\}$ (with $\{f^n\}$ understood) saying, for example, $\{b^n\}$ is a basis, or X has AK iff $\{\delta^n\}$ is a basis.

3. DEFINITION. *We say that* $\{b^n\}$ *is a Toeplitz basis for* X *if there exists a regular matrix* T *such that* $\sum f_k(x)b^k$ *is* T*-summable to* x *for every* $x \in X$.

This means that if we set $y^k = \sum_{i=1}^{k} f_i(x)b^i$, then $\sum t_{nk}y^k \to x$ in X. *A basis is a Toeplitz basis* since we may take $T = I$ in that case.

4. EXAMPLE. The set of trigonometric functions (and Fourier coefficients) forms a Toeplitz basis, with $T = (C,1)$, and not a basis, for the (periodic 2π) continuous functions on $[-\pi,\pi]$ with the topology of uniform convergence [74A] 13.33, p. 441 #15, [79] 7.6 Application 3, and for $L[-\pi,\pi]$ with almost everywhere convergence [74] 13.34.

5. EXAMPLE. Let $(Ax)_n = x_n + x_{n-1}$. Then $\{\delta^n\}$ *is a Toeplitz basis for* c_A^o, hence $\{1,\delta^n\}$ is a Toeplitz basis for c_A. (But not a basis: see 13.3.1 (iv)', 13.4.10) The (C,1) matrix works: Let $h_n(x) = x - (1/n)\sum_{k=1}^{n} x^{(k)}$. Consulting 11.1.6, we see that for $f \in c_A'$, $\lim f[h_n(x)]$ exists and so for each x, $\{h_n(x)\}$ is weakly bounded, hence bounded (8.0.2). By the uniform boundedness principle (7.0.2) $\{h_n\}$ is equicontinuous (= norm bounded in this case.) Now $h_n(x) \to 0$ for $x \in \phi$, hence also for $x \in \overline{\phi}$ (7.0.3). But $\overline{\phi} = c_A^o$ (3.3.8). This argument is similar to that used in 10.3.19; the latter is not available here since A is not an AB matrix.

Since the whole gamut of regular matrices is available it

seems at first glance difficult to find an example in which $\{b^n\}$ is not a Toeplitz basis, given the trivially necessary assumption that it is fundamental (3.0.1). However an example is easily deduced from:

6. LEMMA. *If* $(\{b^n\}\{f_n\})$ *is a Toeplitz basis,* $\{f_n\}$ *is total.*

This means that x must be 0 if $f_n(x) = 0$ for all n. The proof is trivial.

7. EXAMPLE. Let $b^n = 1 + \delta^n$, $f_n(x) = x_n - \lim x$ for $x \in c$. This is a biorthogonal system. To see that $\{b^n\}$ *is fundamental,* let $f \in c'$, say $f(x) = \chi \lim x + tx$ (1.0.2), and $f(b^n) = 0$ for all n. Then $\chi + \sum t_i + t_n = 0$ so t_n is constant. Since $t \in \ell$, $t = 0$, so $\chi = 0$, $f = 0$. The result follows by the Hahn-Banach theorem 3.0.1. However $f_n(1) = 0$ so $\{b^n\}$ *is not a Toeplitz basis* by Lemma 6.

Finally we consider the problem of giving examples in which $\{b^n\}$ is fundamental and $\{f^n\}$ is total. In particular, consider $(\{\delta^n\},\{P_n\})$ in an AD space X:

8. LEMMA. *Let* X *be an FK space in which* $\{\delta^n\}$ *is a Toeplitz basis. Then* $F = W$.

Here F and W are the distinguished subspaces defined in 10.2.3, 5.6.1. Let $x \in F$, $f \in X'$. Then $\sum x_k f(\delta^k)$ is convergent by definition of F. Its sum must be $f(x)$ (forcing $x \in W$ and concluding the proof) since there exists a regular matrix T such that $\lim_T x^{(k)} = x$ implying that $\lim_T f(x^{(k)}) = f(x)$.

9. EXAMPLE. *A Banach space with biorthogonal system* $\{b^n\}\{f^n\}$, *with* $\{b_n\}$ *fundamental and* $\{f_n\}$ *total, which is not a Toeplitz basis.* By Lemma 8 it is sufficient to take A to be a coregular

triangle so that $1 \in F\backslash W$ by 10.2.7 and such that c_A has AD. One is shown in 5.2.5. Of course $b^n = \delta^n$, $f_n = P_n$.

10. EXAMPLE. *The Banach space* c *has a fundamental-total biorthogonal system which is not a Toeplitz basis.* For c is equivalent to c_A (Example 9) by the map A. We can write the system out: the b^n are simply the columns of A, the f_n are the rows of A^{-1} i.e. $f_n(y) = (By)_n$ where $B = A^{-1}$.

19.3. MISCELLANY.

1. There are of course many ways to define generalized limits other than by means of matrices. For example, methods associated with Abel and Borel are well known; others of a very general nature have been introduced by L. Wlodarski and by A. Persson. We refer to [88].

2. One can also extend lim from c to ℓ^∞ by the Hahn-Banach theorem and obtain a generalized limit for bounded sequences. This cannot be given by a matrix since such a matrix would be regular and coercive contradicting 3.5.5. A special case is the Banach limit with the additional property $L\{x_n\} = L\{x_{n+1}\}$. [80] #2-3-106.

3. M. Henriksen [35] introduced another sort of limit on ℓ^∞ which cannot be a Banach limit ([80] #9-6-103); namely let $t \in \beta N\backslash N$ where βN is the Stone-Cech compactification of the positive integers and define $L(x)$ to be the unique value at t of x, extended as a continuous function on βN. This idea led to a fruitful theory; we refer to [1A], [35], [64], [65], [66].

4. In a recent conversation Jeff Connor told me an elegant proof of the fact that no regular matrix can sum every member of

2^N, the set of all sequences of 0's and 1's. (Of course 3.5.5 is an immediate consequence.) The space 2^N, with the product topology (the relative topology of ω) is a complete metric space. If it is included in c_A with A regular, $\lim_A$ is a function of Baire class 1 since it is $\lim(Ax)_n$, hence is continuous at some point [44A], Ch 2, §27, Théorème 1. But every point is the limit of two sequences, one of members of ϕ, the other of points of the form $1 + x$, $x \in \phi$. This proof shows that the generalized limits mentioned above are not of Baire class 1 — and probably can be adapted to show that they are not of any Baire class by Baire's continuity theorem [44A], Ch 2, §28.

5. Let A be a regular matrix and X a locally convex space. It is easy to prove that if $x^n \to 0$ then $\sum_k a_{nk}x^k \to 0$. However, let $X = \ell^{1/2}$, $x^n = \delta^n/n$. Then $\|(1/n)\sum_{k=1}^n x^k\|_{1/2} = n^{-1/2}\sum_{k=1}^n k^{-1/2} > n^{-1/2}\cdot n\cdot n^{-1/2} = 1$. Thus *$\ell^{1/2}$ is not locally convex.*

6. Various sufficient conditions for a space to be conull are given in 6.4.3, 9.3.6, 16.2.6, 16.2.7. A quotient, or even an isomorphic image of a conull space need not be conull, for example if A is a triangle c_A and c are equivalent. *If A is a vsc matrix with $c_A = \omega_A$, then A must be conull.* If not, say $\chi(A) = 1$. Let $M = A-I$. Then M is conull so there exists $x \in c_M\setminus c$ (9.3.6). Then $Ax = Mx + x \notin c$ so $x \in \omega_A\setminus c_A$.

19.4. APPLICATIONS

The first applications were possibly mystical in nature — to attach a "limit" to a divergent series as mentioned in §1.1. In the last century these were used for analytic continuation — for example $\sum z^n$ diverges for $|z| > 1$ but is $1/(1-z)$ for $|z| < 1$; one can declare this to be true for $z = -1$ and get $\sum (-1)^n = \frac{1}{2}$. This gives a summability method, but the practical procedure was the converse: to define some sort of matrix or integral transform and use it to define a function (given by a series) outside the place where its original series converges. See [88] pp. 145, 188.

Fejer's theorem on (C,1) summability of Fourier series (19.2.4) gives the Weierstrass approximation theorem by approximating each trigonometric function by a segment of its power series expansion. It shows also the totality of these functions: if $\int_{-\pi}^{\pi} f(x)t(x)dx = 0$ for every $t(x) = \cos nx$ or $\sin nx$ then f (if continuous) is 0 since its Fourier series is 0. See also [88], p. 154.

A useful possibility is that, even if $\sum x_n$ is convergent, some matrix transformation will converge more quickly so that computation of the sum becomes feasible. (It is easy to write series whose sum would take inadmissible amounts of computer time to calculate directly.) It is also possible to obtain asymptotic expansions this way. See [25], top of p. 96, [88], p. 146. A recent reference is Math. Reviews May 1983 #40003.

An application of summability due to A. K. Snyder is to showing that if one (ideal) point is removed from the Stone-Cech compactification of the rationals the result is countably compact. See [80A], Table 7, p. 339.

B. E. Rhoades [51], [52] has applications to fixed point theorems and Markov chains.

Classical theorems on convergence and interchange of summation can be obtained: *let* $\sup_{m,n} |\sum_{k=1}^{m} a_{nk}| < \infty$, $\sum_k a_{nk}$ *converge for each* n. *Then for any* $t \in \ell$ *the series* $\sum_k \sum_n t_n a_{nk}$ *converges and is equal to* $\sum_n \sum_k$. The first part is simply 8.4.1B which says that $A^T \in (\ell:cs)$. For the second part let $f(t)$ be the first sum. Then f is continuous on ℓ by the Banach-Steinhaus theorem 1.0.4 and, since ℓ has AK, $f(t) = \sum t_n f(\delta^n)$. Another result of this type is 9.6.5.

An interesting elementary application was given by M. Aissen [1]. By the binomial theorem $(1+1/n)^n = \sum_{k=0}^{n} a_{nk}(1/k!)$ where a_{nk} is easily calculated and seen to be in $(cs:\mathbf{c})$ by 8.4.5B. Moreover $\lim_A x = \sum x_n$ for $x \in cs$. Hence $\lim(1+1/n)^n = \sum 1/k!$.

Applications to the study of harmonic functions and H^p spaces are given in [67], [68].

Inclusion theorems can be used to prove that certain spaces are barrelled. Suppose an FK space $X \supset \ell^{1/2}$. Then by 4.2.4 (which does not need local convexity) $\{\delta^n\}$ is bounded in X. By 8.2.5 $X \supset \ell$. This is sufficient to conclude that $\ell^{1/2}$ is a barrelled subspace of ℓ, [80] Example 15-2-5.

Some impossibility proofs can be proved, for example no sequence $\{a^n\}$ of sequences can be found such that $x \in \ell^\infty$ iff $\sum a_k^n x_k$ converges for each n. This is just the statement that $\ell^\infty \neq \omega_A$ and is true because the latter space has AK (4.3.8). In [75] it is deduced from 4.6.6 and 6.4.3 that the definition of Cauchy sequence cannot be given by a countable set of sentences of a certain type.

19.5. QUESTIONS

As with any subject *dignus suo sale* the number of open questions can be multiplied without limit. Some of these seize the imagination because they are easy to state (1,3) or are deceptively easy looking (4,5). Some are so general that they seem easy to disprove (2). Some seem hopeless (9,10).

We shall list a few problems of special interest (to the author) and refer to others in the text. The first one really has nothing to do with summability.

1. Suppose that an AD space X has $X^{\beta} = \ell$. Must $X = c_o$? (8.6.8, 16.3.10).

2. I conjecture that all matrices are μ-continuous. (14.3)

3. I conjecture that a multiplicative triangle must be μ-unique (15.2.13)

4. Find a type M triangle not in the closure of $G(\Delta)$ (18.2.8).

5. Must all (one to one) non-Tauberian matrices lie in the boundary of the maximal group of Γ? (18.2.10)

6. Must a one to one matrix have a growth sequence? (5.5.6)

7. Must a matrix with closed inset be replaceable? (12.2.21)

8. Does there exist a separable coregular pseudoconull space? (16.2.10)

9. Find conditions on a matrix A equivalent to the property that c_A is a BK space.

10. Characterize spaces of the form c_A among FK spaces. (19.1.4)

11. Characterize A such that c is closed in c_A. (3.4)

Other questions occur in 9.5.2 (converse?), 9.5.10, 10.3 (M(X)), 14.5.2, 14.5.6, 15.5.5, 15.5.9, 15.6.6.

19.6 HISTORY

We have spoken of the seed idea of summability in §1.1: to

"sum" divergent series. Ways of doing this were called summability methods; some are described in Chapter 2. Their number is legion and their invention has stretched over the past 150 years. The index of methods on p. 307 of [88] contains 99 entries. Some motivations for particular methods are indicated in §19.4. In 1911, the first step towards abstraction came with the discovery of conditions for regularity of an unspecified matrix (1.3.6). Thought of as a mapping theorem, this led to the investigation of spaces (X:Y) as in Chapter 8. The first applications of functional analysis were by Mazur in 1927 and Banach in 1932; the first is described in §3.1 and 3.3.6; the second is the proof of 1.3.6. The latter, in particular, attracted great attention by the extreme brevity of its proof in comparison with the classical construction used by Toeplitz 21 years earlier. This is by now an old story; the most classical of real and complex variabilists use uniform boundedness, closed graph and Gelfand representation.

The rest of the history is as described (incompletely) in this book. Triangles in Chapter 3, FK spaces in Chapter 4 allowing matrices of any shape and finally theorems about the FK spaces themselves. See §4.1. A thread running from triangles into more less pure functional analysis is described in [82A].

This history referred to is that of applications of functional analysis. As just mentioned it is far from complete — moreover we have omitted all of the large area of summability in which methods of classical analysis predominate. The reader may consult [88] and the Mathematical Reviews. A word of warning: articles written on the subject matter of this book are usually listed under Functional Analysis.

BIBLIOGRAPHY

Abbreviations

AM Annals of Math.
AMM Amer. Math. Monthly
AMS Amer. Math. Soc.
DMJ Duke Math. Journal
IM Indagationes Math.
JDAM Journal d'Analyse Math.
JFA Journal of Functional Analysis
JLMS Journal of LMS
JRAM Journal für die reine und angew. Math.
LMS London Math. Soc.
MMJ Michigan Math. Journal
MR Math. Reviews
MZ Math. Zeitschrift
PAMS Proceedings AMS
PCPS Proc. Cambridge Phil. Soc.
PJM Pacific Jour. Math.
PLMS Proceedings LMS
SM Studia Math.
TAMS Transactions AMS
TMJ Tokoku Math. Journal

1. M. Aissen: The Toeplitz-Silverman Theorem and the definition of e^x. AMM 67(1960) 70-71.

1A. R.E. Atalla: On the multiplicative bheavior of regular matrices PAMS 26(1970) 437-446.

2. S. Banach: Théorie des opérations linéaires. Hafner 1932.

3. H. Baumann: Quotientsätze für Matrizen. MZ 100(1967) 147-162.

4. W. Beekmann: Mercer-Sätze für AB Matrix transformationen. MZ 97(1967) 154-157.

5. W. Beekmann: Über einige limitierungstheoretische Invarianten. MZ 150(1976) 195-199.

6. W. Beekmann, J. Boos, K. Zeller: Die Teilraum P ist invariant. MZ 130(1973) 287-290.

7. W. Beekmann, S.C. Chang: Some summability invariants. Manuscripta Math. 31(1980) 363-378.

8. G. Bennett: A representation theorem for summability domains PLMS 24(1972) 193-203.

9. G. Bennett: Distinguished subspaces and summability invariants SM 40(1971) 225-234.

10. G. Bennett: A new class of sequence spaces. JRAM 266(1974) 49-75.

11. G. Bennett, N.J. Kalton: Consistency Theorems for almost convergence. TAMS 198(1974) 23-43.

12. G. Bennett, N.J. Kalton: FK spaces containing c_0. DMJ 39(1972) 561-582, 819-821.

13. G. Bennett, J.J. Sember, A. Wilansky: Sections of sequences in matrix domains. Trans. N.Y. Acad. of Sci. 34(1972) 107-112.

14. I.D. Berg: Open sets of convervative matrices PAMS 16(1965) 719-724.

14A. I.D. Berg, A. Wilansky: Periodic, semiperiodic and almost periodic sequences. MMJ 9(1962) 363-368.

15. H.I. Brown, D.R. Kerr, H.H. Stratton: The structure of $B(c)$. PAMS 22(1969) 7-14.

16. J. Boos: Ersetzbarkett von Matrixverfahren SM 51(1974) 71-79.

17. J. Boos: Vertraglichkeit von Matrixverfahren MZ 128(1972) 15-22.

18. J. Boos: Zwei-Normen-Konvergenz und Vergleich von beschränkten Wirkfeldern. MZ 148(1976) 285-294.

19. J.W. Calkin: Two sided ideals in $B(H)$. AM 42(1941) 839-873.

20. F.P. Cass: Subalgebras of $B(\ell)$ and βN. JFA 32(1979) 272-276.

21. F.P. Cass, B.E. Rhoades: Mercerian theorems via spectral theory. PJM 73(1977) 63-71.

22. Y. Censor: General conservative FK spaces and an extended definition of conullity. JDAM 33(1978) 105-120.

22A. S.C. Chang, M.S. MacPhail, A.K. Snyder, A. Wilansky: Consistency and replaceability for conull matrices. MZ 105(1968) 202-212.

22B. M.D. Choi: Tricks or treats with the Hilbert matrix AMM 90(1983) 301-312.

23. J. Copping: Mercerian theorems and inverse transformations. SM 21(1962) 177-194.

24. J. Copping: Inclusion theorems for summation methods. IM 20(1958) 485-499.

25. J.H. Curtiss: Book Review. AMM 72(1965) 94-96.

26. N.A. Davydoff: The sharpness of the Mazur-Orlicz theorem. See MR 46(1973) #5884.

27. J. DeFranza, D.J. Fleming: The Banach algebra of conservative triangular matrices. MZ 180(1982) 463-468.

27A. E.K. Dorff, A. Wilansky: Remarks on summability. JLMS 35(1960) 234-236.

28. N. Dunford, J.T. Schwartz: Linear Operators, Part I. Interscience 1958.

29. D. Franekic: Matrix, almost matrix and Tauberian operators on ℓ.c. spaces. Lehigh University Ph.D. Thesis 1974.

30. D.J.H. Garling: On topological sequence spaces. PCPS 63(1967) 997-1019.

31. D.J.H. Garling: The β and γ duality of sequence spaces. PCPS 63(1967) 963-981.

32. D.J.H. Garling, A. Wilansky: On a summability theorem of Berg, Crawford and Whitley. PCPS 71(1972) 495-497.

33. G. Goes: Summen von FK-Räumen. FAK und Umkehrsätze. TMJ 26(1974) 487-504.

34. G.H. Hardy: Divergent Series. Oxford 1949.

35. M. Henriksen: Multiplicative summability methods and the Stone-Cech compactification. MZ 71(1959) 427-435.

36. R.H. Herman: Uniqueness of ideals of compact and strictly singular operators. SM 29(1967) 161-165; MR 36(1968) #5707.

37. E. Hewitt, K. Stromberg: Real and Abstract Analysis. Springer-Verlag 1965.

38. A. Jakimovski, B.E. Rhoades, J. Tzimbalario: Hausdorff matrices as bounded operators over ℓ^p. MZ 138(1974) 173-181.

39. A. Jakimovski, D.C. Russell, J. Tzimbalario: Inclusion theorems for matrix transformations. JDAM 26(1973) 391-404.

40. N.J. Kalton, A. Wilansky: Tauberian operators on Banach spaces. PAMS 57(1976) 251-255.

40A. E.P. Kelly Jr., D.A. Hogan: Bounded conservative linear operators and the boundary of the maximal group. PAMS 32(1972) 195-200.

41. K. Knopp: Folgenräume und Limitierungsverfahren. Rendiconti di Mat. 11(1952) 1-30.

42. K. Knopp, G.G. Lorentz: Beiträge zur absoluten Limitierung. Arch. Math. 2(1949) 10-16.

43. S-Y. Kuan: Some invariant properties on summability domains PAMS 64(1977) 248-250.

44. S-Y. Kuan: On the inset of a convergence domain. PAMS 71(1978) 241-242.

44A. K. Kuratowski: Topologie. Haffner, New York 1952.

45. B. Kuttner, L.H. Lawrence: On Mercerian sets. JLMS 16(1977) 96-98.

46. M.S. MacPhail, A. Wilansky: Linear functionals and summability invariants. Canad. Math. Bull. 17(1974) 233-242.

47. C. Madden: A representation of c_A'. Lehigh University Ph.D. Thesis 1978.

47A. I.J. Maddox: Elements of Functional Analysis. Cambridge 1970.

47B. I.J. Maddox: Infinite Matrices of Operators. Springer Lecture Notes #786, 1980.

48. W. Miesner: The convergence fields of Nörlund means. PLMS 15(1965) 495-507.

48A. M.R. Parameswaran: On the reciprocal of a k-matrix. J. Indian Math. Soc. 20(1956) 329-331.

49. A. Pietsch: Nuclear Locally Convex Spaces. Springer-Verlag 1969.

50. L.A. Raphael: Matrices mapping the space of analytic sequences into itself. PJM 27(1968) 123-126.

B.E. Rhoades:

51. Fixed point iterations using infinite matrices. I. TAMS 194(1974) 161-176; II. Constructive and Computational Methods for Differential and Integral Equations. Springer-Verlag Lecture Notes 430(1974) 390,294; III. Fixed Points, Algorithms and Applications (Ed: S. Karamardian), Academic Press 1977, pp. 337-347.

52. The convergence of matrix transformations for certain Markov chains. Stochastic Processes and their Applications 9(1979) 85-93.

53. Triangular summability methods and the boundary of the maximal group. MZ 105(1968) 284-290.

54. B.E. Rhoades, A. Wilansky: Some commutants in B(c). PJM 49(1973) 211-217.

55. W.H. Ruckle: Sequence Spaces. Pitman 1981.

56. W.H. Ruckle: FK spaces in which the sequence of coordinate vectors is bounded. CJM 25(1973) 973-978.

57. W.H. Ruckle: Representation and series summability of complete biorthogonal sequences. PJM 34(1970) 511-528.

58. D.C. Russell: Inclusion theorems for section-bounded matrix transformations. MZ 113(1970) 255-265.

59. W.L.C. Sargent: On sectionally bounded BK spaces. MZ 83(1964) 57-66.

60. J.J. Sember: A note on conull FK spaces and variation matrices MZ 108(1968) 1-6.

61. J.J. Sember, M. Raphael: The unrestricted section properties of sequences. CJM 31(1979) 331-336.

62. N.K. Sharma: Spectra of conservative matrices. PAMS 35(1972) 515-518.

62A. N.K. Sharma: Isolated points of the spectra of conservative matrices. PAMS 51(1975) 74-78.

A.K. Snyder:

63. Consistency theory in semiconservative spaces. SM 5(1982) 1-13.

64. Cech compactification and regular matrix summability. DMJ 36(1969) 245-252.

65. Heavy points in countable spaces. JDAM 20(1967) 271-279.

66. Generating heavy points with positive matrices. PAMS 19(1968) 973-975.

67. Harmonic functions and regular matrix summability. IJM 13(1969) 406-413.

68. Sequence spaces and interpolation problems for analytic functions. SM 39(1971) 137-153.

A.K. Snyder, A. Wilansky

69. Inclusion theorems and semiconservative FK spaces. Rocky Mtn. J. of Math. 2(1972) 595-603.

70. The Mazur-Orlicz bounded consistency theorem. PAMS 80(1980) 374-376.

71. Non-replaceable matrices. MZ 129(1972) 21-23.

72. M. Stieglitz, H. Tietz: Matrixtransformationen von Folgenraümen. Eine Ergebnisübersicht. MZ 154(1977), 1-16.

73. A.E. Taylor: Functional Analysis. John Wiley 1958.

74. A.E. Taylor: Theory of Functions and Integration. Blaisdell 1965.

74A. E.C. Tichmarsh. The Theory of Functions. Oxford 1939.

A. Wilansky:

75. The Cauchy criterion. AMM 64(1957) 469-471.

75A. Distinguished subspaces and summability invariants. JDAM 12(1964) 327-350.

76. Topics in Functional Analysis. Springer Lecture Notes #45, 1967.

77. Toplogical divisors of 0. TAMS 113(1964) 240-251.

78. Subalgebras of B(X). PAMS 29(1971) 355-360; 34(1972) 632.

79. Functional Analysis. Blaisdell 1964.

80. Modern Methods in Topological Vector Spaces. McGraw Hill 1978.

80A. Topology for Analysis. R.E. Krieger, Melbourne, Florida 1983.

81. The μ property of FK spaces. Commentationes Math 21(1978) 371-380.

82. Semi-Fredholm maps of FK spaces. MZ 144(1975) 9-12.

82A. From triangles to separated inductive limits. SM 31(1968) 469-479.

A. Wilansky, K. Zeller:
83. The inverse matrix in summability. JLMS 32(1957) 397-408.

84. Abschnittebeschränkte Matrixtransformationen. MZ 64(1956) 258-269.

85. Bounded divergent sequences, topological methods. TAMS 78(1955) 501-509; 80(1955) 386.

85A. Banach algebra and summability. IJM 2(1958) 378-385; 3(1959) 468.

86. K. Zeller: Abschnittskonvergenz in FK-Räumen. MZ 55(1951) 55-70.

87. K. Zeller: Faktorfolgen bei Limitierungsverfahren. MZ 56(1952) 134-151.

88. K. Zeller, W. Beekman: Theorie der Limitierungsverfahren. Springer-Verlag 1970.

89. M. Zippin: The separable extension problem. Israel J. Math. 26(1977) 372-387.

90. A. Zygmund: Two notes on the summability of infinite series. Colloq. Math 1(1947-8) 225-229.

INDEX

SYMBOLS

(These are listed alphabetically in the above index where possible)

NAMES